COUNTRIES
OF THE WORLD
IN MINUTES

JACOB F. FIELD

Quercus

CONTENTS

Introduction	6
World map	8
Europe	10
Africa	104
Asia	214
Australasia	308
Oceania	314
North America	340
South America	388
Index of Countries	414
Acknowledgements	416

Each entry in this book features a map showing key cities (with population information for the capital, starred, and most populous city), and major geographical features (including the world's major rivers and twenty highest mountain ranges). Data boxes provide key information in a compact form:

Official name
Latitude and longitude of capital
Time zone (relative to UTC, Universal Coordinated Time)

Key to country data:

□	Total surface area
O	Climate type
†	Total population
⊡	Population density
††	Population growth rate (annual)
◠	Urbanization – proportion of urban to rural populations
⋔	Ethnic breakdown from census figures (individually listed groups form more than 10 per cent of population)
♥	Dominant language(s)
▢	Dominant religion(s)
▰	Structure of government
◉	Membership of major international organizations – see listing at right.
▧	Currency
▦	GDP (Gross domestic product) – market value of goods and services produced annually.
▨	GDP per head of population
▰	Major national industries and sources of revenue

Key to international organizations:
ACD: Asia Cooperation Dialogue
AL: Arab League
AMU: Arab Maghreb Union
ASEAN: Association of Southeast Asian Nations
AU: African Union
CAN: Andean Community
CAP: Central American Parliament
CARICOM: Caribbean Community
CIS: Commonwealth of Independent States
COE: Council of Europe
CON: Commonwealth of Nations
EAC: East African Community
EAEU: Eurasian Economic Union
ECCAS: Economic Community of Central African States
ECOWAS: Economic Community of West African States
EFTA: European Free Trade Association
EU: European Union
G7: Group of Seven Major Advanced Economies
G-15, G24, G77: Groups and subgroups of developing nations sharing specific policy goals.
GCC: Gulf Cooperation Council
IGAD: Intergovernmental Authority for Development
IMF: International Monetary Fund
Mercosur: Union of South American Nations
NAFTA: North American Free Trade Agreement
NAM: Non-Aligned Movement
NATO: North Atlantic Treaty Organization
OAS: Organization of American States
OECD: Organization for Economic Cooperation and Development
OPEC: Organization of Petroleum-Exporting Countries
PA: Pacific Alliance
PIF: Pacific Islands Forum
SAARC: South Asian Association for Regional Cooperation
SADC: Southern African Development Community
SCO: Shanghai Cooperation Organization
UN: United Nations
WB: World Bank
WTO: World Trade Organization

Introduction

Our world is complex and diverse. If you want to understand our interconnected global society, then you cannot limit your perspective. It is imperative to look at countries from every corner of the planet – from the largest and most prosperous to the smallest and least developed. Have you ever wanted to know which country's name means Land of Incorruptible People? Or which island-nation made (and lost) a fortune out of mining bird droppings? Or when the last oldest existing national constitution in the world was put in place? This book will enlighten you about every country in the world by providing all the key information on each and every one.

Each of the 193 sovereign states that make up the United Nations (as well as the two that have official observer status: Palestine and the Vatican City) has its own individual entry and is given equal coverage in *Countries of the World in Minutes*. This means that no region is left out or underrepresented, which ensures you will gain a more complete understanding of the world. For each nation there is a short overview of the most important features of its history, politics, culture and geography. A data box provides the key facts, such as ethnic breakdown, climate and main industries, as well as a statistical snapshot of economy and population (see the key on page 5 for more information).

This book is structured so that there is a chapter on each inhabited continent (if you need to find a specific country quickly you can just check the index). The first continent to be considered is Europe, location of Coordinated Universal Time (abbreviated as UTC), which acts as a benchmark for all of the other time zones in the world. The book then proceeds eastwards, covering Africa, Asia, Australasia, Oceania, North America and South America. Within each chapter countries are arranged in geographical order by their westernmost point – of the mainland if they have overseas territories; this means that nations covering more than one continent, such as Egypt and Russia, are placed in Africa and Europe, respectively.

Whenever any country, no matter where it is, comes up in the news, *Countries of the World in Minutes* will allow you to gain an instant insight into it.

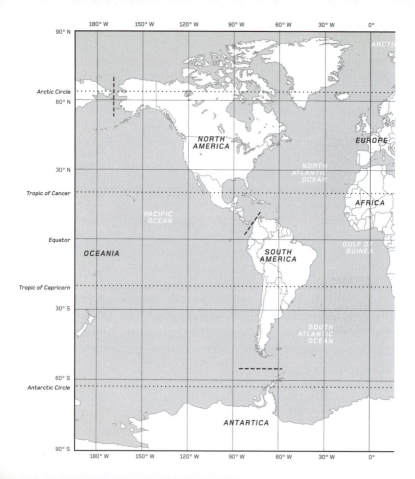

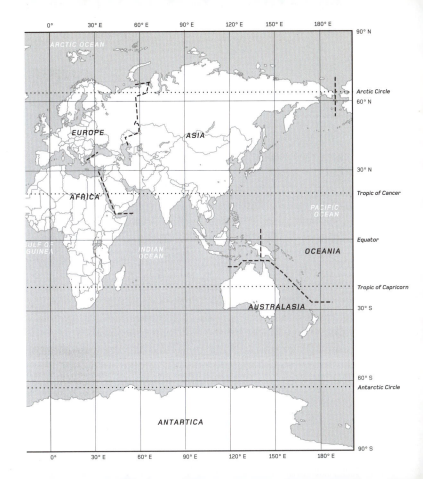

Europe

The western portion of the Eurasian land mass, Europe is bounded to the south by the Mediterranean Sea, to the west by the Atlantic Ocean and to the north by the Arctic. Its border with Asia is marked by the Black and Caspian seas and the Ural and Caucasus mountains. Humans arrived here via Asia about 40,000 years ago. Since then, Europe has been the centre of many empires, including those of Alexander the Great and Rome, as well as more modern regimes established by Britain, Russia, Spain, France and Portugal. Western Europe was the first area in the world to experience the Industrial Revolution, transforming its economy and society from the 1760s onwards. In the 20th century, rivalries between European powers provoked two world wars (1914–18 and 1939–45). Since 1945, Europe has become more peaceful and integrated, with the major driving force of this unity being the European Economic Community (EEC), created in 1957 and precursor to the European Union (EU) established in 1993. Europe is an economic powerhouse, accounting for over ten per cent of the world's population, but around one-third of its GDP.

Iceland

Straddling the Mid-Atlantic Ridge that divides the Eurasian and North American tectonic plates, Iceland is subject to earthquakes, volcanoes and geothermal activity (the latter, combined with hydroelectric power, generates almost all of Iceland's electricity). Permanent human habitation dates back to the late-ninth century CE, when Scandinavian and Celtic settlers arrived. Iceland became an independent state in 930, governed by an assembly called the Althing, the world's oldest legislature. From 1262, Iceland became subject to Norwegian, and subsequently Danish, rule. The country won limited self-government in 1874, but it took another 70 years to achieve full independence. Once one of the poorest countries in western Europe, Iceland developed rapidly from the mid-20th century, thanks largely to its fishing industry. From 2003, Icelandic banks became increasingly active on the international stage, where risky investments led to a major financial crisis from 2008–11. Despite this, Icelanders enjoy extremely high standards of living and social equality.

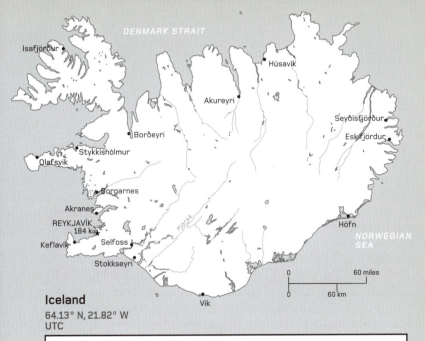

DENMARK STRAIT

Ísafjörður

Húsavík

Akureyri

Seyðisfjörður

Borðeyri

Eskifjörður

Stykkishólmur

Ólafsvík

Borgarnes

Akranes

REYKJAVÍK
184 k

Höfn

Keflavík

Selfoss

NORWEGIAN
SEA

Stokkseyri

Þjórsá

Vík

| 0 | | 60 miles |
| 0 | | 60 km |

Iceland

64.13° N, 21.82° W
UTC

☐ 103,000 km² (39,769 sq mi)	⚤ 94% Icelandic 6% others	🖳 Icelandic Króna (ISK)
○ Temperate	👤 Icelandic	📊 $20.05 bn
⚤ 334,252	📖 Christianty	📊 $59,977
⊞ 3.3/km² (8.5/sq mi)	⚖ Unitary parliamentary republic	🖪 Fishing, tourism geothermal/hydro power
⚤⚤ 1.0%	⚙ COE, EFTA, IMF, NATO, OECD, UN, WB, WTO	
☐ 94.2 : 5.8%		

Ireland

A country of central plains ringed by coastal highlands, the Irish landscape features many lakes and rivers. Once a collection of earldoms, Ireland was joined to Great Britain in 1801, to form the United Kingdom. A century later, demands for self-rule culminated in warfare between mostly Catholic Irish nationalists and the British. This led to self-government in 1922, although six Protestant-majority counties remained part of the UK, as Northern Ireland. Ireland declared its independence from Britain in 1937, and became a republic in 1949.

Reliant on agriculture for centuries, the Irish have known much poverty; the Great Famine of the 1840s led to one-third of the country's population dying or emigrating. After joining the EEC in 1973, Ireland began reforms that diversified and modernized its economy, enjoying a sustained economic boom from 1995–2007. Although the global crash of 2008 brought this to an end, the Irish economy has since rebounded, partly through incentives for multinational companies, such as Google and Apple, to invest.

NORTH ATLANTIC OCEAN

Finn

NORTHERN IRELAND (U.K.)

Sligo

Annagh

Carrick-on-Shannon

Dundalk

IRISH SEA

Roscommon

Inny

Boyne

Mullingar

DUBLIN 1.2 m ★

Galway

Tullamore

Liffey

Shannon

Barrow

Wicklow

Suir

Kilkenny

Tralee

Suir

Rosslare Harbour

Blackwater

Waterford

Lee

Cork

NORTH ATLANTIC OCEAN

| 0 | | 60 miles |
| 0 | | 60 km |

Republic of Ireland
53.35° N, 6.26° W
UTC

- ☐ 70,280 km² (27,135 sq mi)
- ○ Temperate maritime
- 4,773,095
- 69.3/km² (180/sq mi)
- 2.0%
- 63.5 : 36.5 %
- 85% Irish, 15% others
- English, Irish
- Christianity
- Unitary parliamentary republic
- COE, EU, IMF, UN, WB, WTO
- Euro (EUR)
- $294.05 bn
- $61,606
- Pharmaceuticals and chemicals, computer manufacture, agriculture and food processing

Portugal

Located on the Atlantic coast of the Iberian Peninsula, Portugal became an independent kingdom in 1139. During the 15th and 16th centuries, it grew into a great maritime power, with an empire that extended from Brazil to Southeast Asia. The country fell under Spanish rule in 1580, but regained its independence in 1640, from which time Portuguese prestige and power waned. In 1910, a revolution brought an end to the monarchy, but the subsequent democratic-republican regime was unstable. The one-party authoritarian system of the *Estado Novo* (New State) was established in 1933 and it was not until 1974 that the Carnation Revolution restored democracy.

Most of Portugal's population is concentrated in coastal areas, with the mountainous interior more sparsely populated. The Portuguese economy grew during the 20th century, improving living standards, and in 1986 Portugal joined the EEC. After 2001, Portugal's economy contracted, and the country was badly hit by the global crash of 2008.

Portuguese Republic

38.72° N, 9.14° W
UTC

☐	92,225 km² (35,608 sq mi)
○	Maritime temperate
♀	10,324,611
▥	112.7/km² (292/sq mi)
♀♀	-0.3%
◔	64.0 : 36.0%
♀♀	98% Portuguese, 2% others
♟	Portuguese
▣	Christianity
▥	Unitary semi-presidential republic
◉	COE, EU, IMF, NATO, OECD, UN, WB, WTO
⌦	Euro (EUR)
▦	$204.56 bn
▦	$19,813
▦	Textiles, agriculture, forestry/cork

Spain

The main feature of the Spanish landscape is the Meseta Central, a vast plateau half a mile above sea level that is ringed by mountains, desert and coastal plains. Spain's origins as a unified country began in 1469, with the marriage of the monarchs of Castile and Aragon, the region's two largest kingdoms. During the 16th and 17th centuries, Spain boasted a vast global empire covering most of Central and South America, as well as the Philippines and parts of Italy and the Low Countries. In 1873, with imperial power in decline, the monarchy was briefly abolished, but it returned from 1874-1931 before a republican regime was installed. The Spanish Civil War (1936–39) saw Francisco Franco's nationalist forces defeat the republicans and establish a totalitarian dictatorship that lasted until Franco's death in 1975. After the country transitioned into a democratic constitutional monarchy and joined the EEC in 1986, its economy rapidly modernized. Under the 1978 constitution, Spain has 17 'autonomous communities' with varying degrees of domestic self-government. These different regions retain many of their own customs, laws and languages.

BAY OF BISCAY

Santiago de Compostela
Oviedo
Gijón
Santander
Bilbao
San Sebastián
Vitoria
Pontevedra
Vigo
Orense
Río Miño
Río Esla

FRANCE

ANDORRA

Girona
Zaragoza
Barcelona
Tarragona
Río Ebro

Valladolid
Río Duero
Salamanca

PORTUGAL

MADRID
6.2 m ★

Río Tormes
Toledo

Tagus

Castellón de la Plana
Valencia

BALEARIC SEA

MINORCA
Palma
BALEARIC ISLANDS
MAJORCA
IBIZA

Badajoz
Río Jucar

Río Segura
Alicante
Murcia
Cartagena

Córdoba
Río Genil
Huelva
Sevilla
Granada
Málaga
Almería

NORTH ATLANTIC OCEAN

Cádiz
Gibraltar (U.K.)

GOLFO DE CÁDIZ

ALBORAN SEA

MEDITERRANEAN SEA

MOROCCO

0 — 100 miles
0 — 100 km

There are two small Spanish exclaves in North Africa: Ceuta and Melilla

Kingdom of Spain
40.42° N, 3.70° W
UTC + 1

☐ 549,087 km² (212,004 sq mi)	⌂ Christianity
○ Temperate	♣ Unitary parliamentary system under constitutional monarchy
⚥ 46,443,959	
⊞ 92.8/km² (241/sq mi)	⊕ COE, EU, IMF, NATO, OECD, UN, WB, WTO
⇅ -0.01%	
◻ 79.8 : 20.2%	⬡ Euro (EUR)
ⅲ 88% Spanish, 12% others	▦ $ 1.23 tn
⚑ Spanish (although Catalan, Galician, Basque and Occitan recognized in autonomous communities)	▩ $26,528
	⚒ Machinery and motor vehicles, metals, chemicals

United Kingdom

The island of Great Britain lies within an archipelago off the northwestern coast of continental Europe. Originally a series of smaller kingdoms, the country began to coalesce when England extended control over Wales in the 16th century. The English and Scottish crowns were joined under a single monarch in 1603, although the two kingdoms did not unify until 1707. Britain formally united with Ireland in 1801 to form the United Kingdom. Aside from a brief republican period in the mid-17th century, monarchs have reigned over UK territories for centuries, although their power has steadily diminished and parliament is now the supreme legal authority. Britain led the Industrial Revolution of the late 1700s, and flourished as the world's greatest economic power for much of the 19th century, building a global empire that covered nearly one-quarter of the world's land area. Global power declined during the 20th century, despite victory in both world wars, and the majority of Britain's territories gained independence. Even so, the United Kingdom remains one of the wealthiest and most populous nations in Europe, and is a centre of global finance and banking.

United Kingdom of Great Britain and Northern Ireland

51.51° N, 0.13° W
UTC

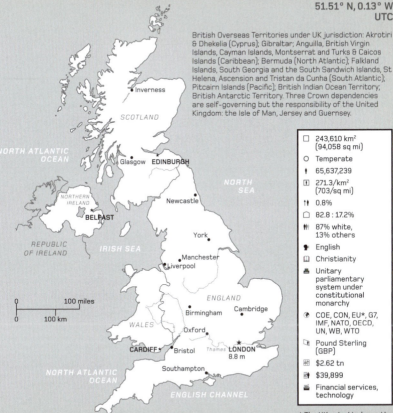

British Overseas Territories under UK jurisdiction: Akrotiri & Dhekelia (Cyprus); Gibraltar; Anguilla, British Virgin Islands, Cayman Islands, Montserrat and Turks & Caicos Islands (Caribbean); Bermuda (North Atlantic); Falkland Islands, South Georgia and the South Sandwich Islands, St Helena, Ascension and Tristan da Cunha (South Atlantic); Pitcairn Islands (Pacific); British Indian Ocean Territory; British Antarctic Territory. Three Crown dependencies are self-governing but the responsibility of the United Kingdom: the Isle of Man, Jersey and Guernsey.

- NORTH ATLANTIC OCEAN
- SCOTLAND
- Inverness
- Glasgow
- EDINBURGH
- NORTHERN IRELAND
- BELFAST
- REPUBLIC OF IRELAND
- IRISH SEA
- NORTH SEA
- Newcastle
- York
- Manchester
- Liverpool
- ENGLAND
- Birmingham
- Cambridge
- WALES
- Oxford
- CARDIFF
- Bristol
- Thames
- LONDON 8.8 m
- Southampton
- NORTH ATLANTIC OCEAN
- ENGLISH CHANNEL

0 — 100 miles
0 — 100 km

- ☐ 243,610 km² (94,058 sq mi)
- ○ Temperate
- ♀ 65,637,239
- ▥ 271.3/km² (703/sq mi)
- ⇈ 0.8%
- ◻ 82.8 : 17.2%
- ⚥ 87% white, 13% others
- ☿ English
- ▢ Christianity
- ⚒ Unitary parliamentary system under constitutional monarchy
- ◉ COE, CON, EU*, G7, IMF, NATO, OECD, UN, WB, WTO
- ▨ Pound Sterling (GBP)
- ▦ $2.62 tn
- ▧ $39,899
- ▬ Financial services, technology

* The UK voted to leave the EU in a 2016 referendum

France

The largest country in western Europe, France has a landscape of plains and hills, with mountainous areas to the south and east. Its modern history began with the French Revolution of 1789, which overthrew a monarchy that had ruled the Kingdom of Francia since the late fifth century, gradually extending its territory across the present-day country. A republic was declared in 1792, but the upheavals that followed saw the country alternate between republican, monarchic and imperial rule. France was particularly influential from 1804–14, when Napoleon Bonaparte extended imperial dominance over much of Europe; through the rest of the 19th century, it built an overseas empire covering parts of South America, Africa and Southeast Asia. Since 1870, it has been a republic with an elected head of state. France was heavily involved in the two world wars, during which it was partly occupied by Germany. Today, it is Europe's leading agricultural producer, principally of wheat, meat and dairy products, and a significant industrial power. The country has been a central force behind the EU, and a major supporter of European integration.

French Republic

48.86° N, 2.35° E
UTC + 1

'Overseas departments': French Guiana (South America); Guadeloupe and Martinique (Caribbean); Mayotte and Réunion (Indian Ocean). Overseas collectivities: French Polynesia and Wallis & Futuna (Pacific Ocean); St Barthélemy and St Martin (Caribbean); St Pierre & Miquelon (North Atlantic). New Caledonia in the Pacific Ocean is a special collectivity. Adélie Land in the Antarctic is also claimed by France.

- □ 547,557 km² (211,413 sq mi)
- ○ Ranges from oceanic to continental to Mediterranean
- ♦ 66,896,109
- ⊡ 122.2/km² (317/sq mi)
- ↑↑ 0.4%
- ⌂ 79.75 : 20.25%
- ♯ 77% French, 23% others
- 🗣 French
- ⛪ Christianity
- ⚖ Unitary senatorial semi-presidential republic
- ☺ COE, EU, G7, IMF, NATO, OECD, UN, WB, WTO
- 💶 Euro (EUR)
- 📈 $2.46 tn
- 💵 $36,855
- 🚂 Machinery and equipment, aircraft and automotive manufacture, agriculture and food processing

Map labels

ENGLISH CHANNEL
BELGIUM
LUXEMBOURG
GERMANY
SWITZERLAND
ITALY
BAY OF BISCAY
MEDITERRANEAN SEA
ANDORRA
SPAIN
CORSICA

Dunkerque, Calais, Lille, Valenciennes, Amiens, Cherbourg, Le Havre, Rouen, Caen, Châlons-sur-Marne, Metz, Nancy, Strasbourg, Brest, PARIS 10.8 m, Troyes, Rennes, Le Mans, Orléans, Lorient, Saint-Nazaire, Tours, Nantes, Poitiers, Dijon, Besançon, La Rochelle, Clermont-Ferrand, Lyon, Limoges, Grenoble, Bordeaux, Toulouse, Montpellier, Nice, Cannes, Marseille, Toulon, Ajaccio

Seine, Meuse, Moselle, Rhine, Marne, Loire, Saône, Doubs, Rhône, Dordogne, Lot, Garonne

ALPS

0 — 100 miles
0 — 100 km

Andorra

Nestled in the Pyrenees mountains, Andorra shares borders with France and Spain. In 1278, a territorial dispute led to Andorran sovereignty being split between the French Count of Foix and the Spanish Bishop of Urgell. In 1607, the French claim passed to the king, who jointly ruled the country with the Bishop of Urgell as a 'co-prince of Andorra'. The tradition continued despite the abolition of the French monarchy; today the French president assumes the title. Direct 'diarchic' rule ended in 1993, with the introduction of a democratic parliamentary system. Andorra now elects a head of government, although the co-princes remain the titular heads of state.

With a territory that is mostly mountainous, Andorra has traditionally relied on pastoral farming, primarily sheep and goat herding. In recent years, its economy has become more focused on retail, offering duty-free shopping to tourists and visitors. The country's tourism sector has also developed, and Andorran ski slopes and resorts are some of the most popular in the Pyrenees.

FRANCE

El Serrat

Llorts

Arinsal

Canillo

Soldeu

SPAIN

Ordino

Pal

La Massana

FRANCE

Anyós

Encamp

ANDORRA
LA VELLA
23 k

Els Vilars d'Engordany

Les Escaldes

Santa Coloma

SPAIN

St. Julià de Lòria

SPAIN

0 ____ 4 miles
0 ____ 4 km

Principality of Andorra
42.51° N, 1.52° E
UTC + 1

☐	470 km² (181 sq mi)	⚒	Unitary parliamentary semi-elective diarchy
O	Temperate	⊕	COE, G77, UN, WTO (observer)
♦	77,281	⌧	Euro (EUR) (via monetary agreement; not a member of the eurozone)
⊡	164.4/km² (427/sq mi)		
♦♦	-0.9%	🗠	$3.24 bn (2013)
◠	84.6 : 15.4%	🗠	$40,215 (2013)
♦♦♦	46% Andorran, 26% Spanish, 13% Portuguese, 15% others	🛆	Tourism
♦	Catalan		
☐	Christianity		

Belgium

The region that is now Belgium experienced periods under Spanish rule (1581–1714) and the Austrian Habsburgs (1714–97). Following the Napoleonic era, during which it was under French rule, it became part of the Kingdom of the Netherlands. The country declared independence in 1830, after which it became a parliamentary democracy with a hereditary monarch as head of state. It was occupied by Germany during both world wars.

Today, Belgium is divided into three regions: Dutch-speaking Flanders in the north; French-speaking Wallonia in the south; and the officially bilingual Brussels-Capital Region. The early 19th century saw rapid industrialization and the establishment of a colonial empire in Africa, which gained independence in the 1960s. Belgium was one of the six founding members of the EEC, and today its capital is home to the headquarters of both NATO and the EU. Disagreements between the Dutch and French populations led to a series of constitutional reforms between 1970 and 1993 that effectively created a federal system.

NORTH SEA

NETHERLANDS

GERMANY

Knokke-Heist

Oostende

Bruges

Antwerp

Gent

Leie

Schelde

Leuven

Hasselt

Kortrijk

Schelde

★ BRUSSELS
1.2 m

Liège

Mons

Namur

Meuse

Malmé-Hdy

Charleroi

Meuse

FRANCE

Bastogne

Kingdom of Belgium
50.85° N, 4.35° E
UTC + 1

LUXEMBOURG

Arlon

☐	30,530 km² (11,788 sq mi)
○	Temperate
♁	11,348,159
⊞	374.8/km² (971/sq mi)
↟↟	0.7%
◠	97.9 : 2.1%
⋔	58% Flemish, 31% Walloon, 11% others
♗	Dutch, French, German

▥	Christianity
⚖	Federal parliamentary system under constitutional monarchy
⊕	COE, EU, IMF, NATO, OECD, UN, WB, WTO
▱	Euro (EUR)
▨	$466.37 bn
▩	$41,096
⛭	Engineering and metals, transportation and automotive assembly

0 — 30 miles

0 — 30 km

Netherlands

L ocated on Europe's northwestern coast, more than half of the Netherlands is less than 1 m (40 in) above sea level. From the late 15th century, the Habsburg dynasty (rulers of Spain from 1516) extended control over the region's previously independent cities, counties and duchies. The Dutch rebelled against their Spanish overlords, but it took an 80-year war for them to establish their independence. During this time (1568–1648), they built a global empire based on maritime commerce and trade. A period of French domination during the French Revolutionary and Napoleonic era (1792–1815) ended with the Dutch reasserting independence. Parliamentary democracy was established in 1848, although the monarch remained head of state. During the Second World War, Germany occupied the Netherlands and it suffered heavy bombings and famine. Today, it has a well-developed economy, with high earnings from agricultural exports driven by the lucrative floral market. A truly egalitarian society, the Netherlands is, perhaps, the most socially liberal country in the world. In 2001, it became the first nation to legalize same-sex marriage.

Netherlands

52.37° N, 4.90° E
UTC + 1

The Caribbean countries of Aruba, Curaçao and Sint Maarten are part of the Kingdom of the Netherlands, with domestic self-government but foreign policy and defence governed by the Netherlands. There are also three Dutch 'special municipalities' in the Caribbean: Bonaire, Sint Eustastius and Saba.

☐	41,540 km² (16,039 sq mi)
○	Temperate
⚲	17,018,408
▦	505.1/km² (1,309/sq mi)
⚤	0.5%
◖	91.0 : 9.0%
⚥	79% Dutch, 21% others
☻	Dutch
☐	Christianity
⚒	Unitary parliamentary system under constitutional monarchy
◉	COE, EU, IMF, NATO, OECD, UN, WB, WTO
⬒	Euro (EUR)
▦	$770.85 bn
▥	$45,295
⚏	Chemicals, petroleum refining, electronics, agriculture and food processing

WEST FRISIAN ISLANDS

NORTH SEA

Leeuwarden

Groningen

Assen

Zwolle

IJssel

GERMANY

Enschede

AMSTERDAM
1.1 m

The Hague

Nederrijn

Arnhem

Rotterdam

Maas

Waal

Nijmegen

's Hertogenbosch

Breda

Eindhoven

GERMANY

BELGIUM

Maas

0   30 miles

0   30 km

Maastricht

Norway

From the eighth to tenth centuries, Norse Vikings raided and traded across Europe. They founded the Kingdom of Norway in 872 – originally centred on the southwestern part of the Scandinavian Peninsula – and joined with Denmark and Sweden in the Kalmar Union, which lasted from 1397 to 1523 when Sweden split from the union. In 1814, the King of Denmark and Norway ceded his rule over the country to the Swedish king. Demands for independence grew, and in a 1905 referendum, the people voted to remain a monarchy and offer the crown to a Danish prince whose ancestors continue to reign; the king is head of state, while an elected prime minister serves as head of government. Despite a declaration of neutrality in the Second World War, the Germans occupied Norway from 1940–45. During the 1960s, the Norwegian economy was transformed when large reserves of oil and gas were discovered in its waters, and the country has since become a highly prosperous and developed nation. Norway is not part of the EU, having rejected membership in two referendums in 1972 and 1984.

Kingdom of Norway

59.91° N, 10.75° E
UTC + 1

Peter I Island and Queen Maud
Land in the Antarctic are
claimed by Norway

HAMMERFEST
Vadsø
Tromsø
RUSSIA
FINLAND
Harstad
VESTERÅLEN
LOFOTEN
Narvik
Bodø
Mo i Rana
Sandnessjøen
Mosjøen
NORWEGIAN
SEA
Namsos
Steinkjer
Trondheim
Molde
Ålesund
SWEDEN
GULF OF
BOTHNIA
Bergen
Lågen
Lågen
Otra
OSLO
986 k
Stavanger
NORTH
SEA
Kristiansand

QUEEN MOUNTAINS

0   100 miles
0   100 km

▢	385,178 km² (148,718 sq mi)
◯	Temperate along coast, colder interior
⛫	5,232,929
⊞	14.3/km² (37/sq mi)
⛫⛫	0.9%
◻	80.7 : 19.3%
⛫⛫	93% Norwegian, 7% others
⛫	Norwegian
⛫	Christianity
⛫	Unitary parliamentary system under constitutional monarchy
◉	COE, EFTA, IMF, NATO, OECD, UN, WB, WTO
⛫	Norwegian Krone (NOK)
▦	$370.56 bn
▦	$70,812
⛫	Oil and gas

Luxembourg

Located between France, Germany and the Low Countries (Belgium and the Netherlands), Luxembourg occupies an important strategic position. Founded in 963 CE, the territory came under Habsburg rule from 1477 until 1795, when it was conquered and annexed by France. In 1815, the country was elevated to the status of Grand Duchy in 1815 and gained independence. The grand duke still acts as the head of state, while political leadership is held by an elected prime minister.

Mindful of a history dominated by rivalries between Europe's great powers, Luxembourg became a proponent of European integration and unity in the second half of the 20th century; it was a founding member of NATO, the EEC and the eurozone (a monetary union that shares the euro currency). Despite being one of the world's smallest states, it is one of the wealthiest per capita. Banking plays a major role in the economy, with the financial sector making up over one-third of GDP, although in recent decades it has diversified into high-tech industries.

Grand Duchy of Luxembourg

49.61° N, 6.13° E
UTC + 1

BELGIUM

Wemperhardt

Clervaux

GERMANY

Vianden

Sauer

Hiederscheid

Ettelbruck

Echternach

Mersch

Grevenmacher

BELGIUM

Steinfort

★ LUXEMBOURG
107 k

Hesperange

Pétange

Remich

Differdange

Dudelange

FRANCE

- ☐ 2,590 km² (1,000 sq mi)
- ○ Modified continental
- ♦ 582,972
- ⊞ 225.1/km² (583/sq mi)
- ↑↑ 2.3%
- ◠ 90.4 : 9.6%
- ♔♔ 53% Luxembourger, 16% Portuguese, 31% Others
- ♟ Luxembourgish, German, French
- ▢ Christianity
- ♞ Unitary parliamentary system under constitutional monarchy
- ⊕ COE, EU, IMF, NATO, OECD, UN, WB, WTO
- ⬓ Euro (EUR)
- ▤ $59.95 bn
- ▦ $102,831
- ⬒ Banking and financial services, steel production

0 ——— 10 miles

0 ——— 10 km

Germany

Extending from the Alps to the North and Baltic seas, the German landscape comprises plains, rivers, forests and mountains that contain significant natural resources, including metals, coal and natural gas. For a millennium the region was part of the Holy Roman Empire, which united hundreds of smaller semi-independent states from 800 to 1806. However, this did not prevent the German lands being torn apart in the brutal Thirty Years' War (1618–48). In 1871, the country was unified under the leadership of Prussia, leading to the creation of an empire (although this collapsed in 1918 following defeat in the First World War). A short period as a parliamentary republic ended when the Nazi Party seized power in 1933. The Nazis led Germany into the Second World War, which left it in ruins and subsequently divided into two states: democratic West Germany (a founding member of the EEC), and socialist East Germany. The two states reunified in 1990 and Germany is now a leading world power, with the largest economy in Europe built on the export of manufactured goods, such as vehicles and machinery.

Federal Republic of Germany
52.52° N, 13.41° E
UTC + 1

- ☐ 357,380 km² (137,985 sq mi)
- ○ Temperate and marine
- ✝ 82,667,685
- ⊡ 236.9/km² (614/sq mi)
- ♕ 1.2%
- ◖ 75.5 : 24.5%
- ♛ 92% German, 8% others
- ♘ German
- ⬚ Christianity
- ⚒ Federal constitutional parliamentary republic
- ✪ COE, EU, G7, IMF, NATO, OECD, UN, WB, WTO
- ⬚ Euro (EUR)
- ▧ $3.47 tn
- ▨ $41,936
- ⬛ Automotive and machinery manufacture, metals, chemicals, electronics

Switzerland

The Swiss Confederation originated in the late 13th century, when three areas (called 'cantons') in Switzerland's central region allied together. Over the centuries, the number of cantons has steadily increased to its current figure of 26, united under a federal republic. Each Swiss canton has a high degree of autonomy, with its own parliament and legal system, while a long tradition of direct democracy continues to see frequent referendums at local, regional and national levels.

Switzerland adheres to a policy of armed neutrality; although it has not been at war since 1815, the country retains a strong military. All sides respected Swiss neutrality in the two world wars, allowing the country to escape the mass bloodshed of these conflicts. Owing to its desire to remain neutral, Switzerland did not join the UN until 2002 and has stayed outside of the EU. One of the wealthiest countries in the world, Switzerland's economy is famed for its stability, prosperity and innovation; it is a major centre of international banking.

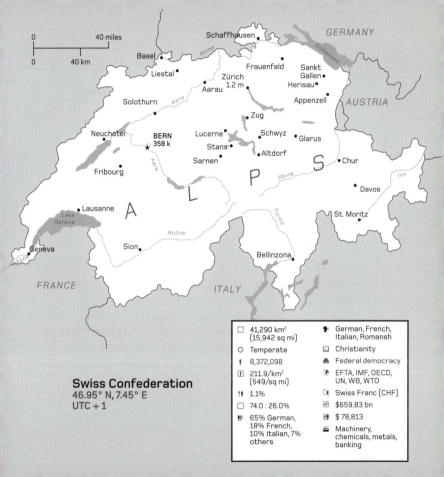

Swiss Confederation
46.95° N, 7.45° E
UTC + 1

- ▢ 41,290 km² (15,942 sq mi)
- ○ Temperate
- ⬙ 8,372,098
- ⊞ 211.9/km² (549/sq mi)
- ⚤ 1.1%
- ▱ 74.0 : 26.0%
- ⚥ 65% German, 18% French, 10% Italian, 7% others

- ❦ German, French, Italian, Romansh
- 📖 Christianity
- ⛪ Federal democracy
- ⊕ EFTA, IMF, OECD, UN, WB, WTO
- 💷 Swiss Franc (CHF)
- ▦ $ 659.83 bn
- ▤ $ 78,813
- ▭ Machinery, chemicals, metals, banking

Italy

Once the seat of the Roman Empire that ruled much of ancient Europe, North Africa and the Near East, Italy fragmented into rival states following the Empire's collapse in the fifth century CE. The region has been a major cultural influence in Europe, particularly during the Renaissance, which flourished in the city-states of Florence, Venice, Milan and Rome from the 14th to 17th centuries.

Unification was achieved in 1861, after which Italy industrialized rapidly, particularly in the north. A democratic constitutional monarchy ended in 1922 when the National Fascist Party seized power, instituting a totalitarian dictatorship that allied with Germany during the Second World War. In 1946, the country voted to become a republic and abolished the monarchy. A postwar revival brought sustained economic recovery, during which the country became a founding member of NATO, the EEC and the eurozone. Investigations in the early 1990s exposed links between politics, business and organized crime, and corruption remains a problem; government and courts continue their attempts to combat it.

Monaco

Located on the Mediterranean Sea, Monaco's land border is shared entirely with France. In 1191, the Holy Roman Empire granted the Italian city-state of Genoa control of this area, where they built a fortress in 1215. The Grimaldi, an Italian noble house, seized control of the territory in 1297 and has ruled it (with occasional interruptions) ever since. Monaco became a French protectorate in 1641, and it was not until 1861 that Monégasque independence and sovereignty were fully and formally recognized. The country rose to fame as a leisure destination for the wealthy from the late 19th century, having built its first casino in Monte Carlo. It also benefits from an advantageous financial regime, with no income tax for residents and a banking system famed for its strict confidentiality. Monaco was an absolute monarchy until 1911 and the adoption of a constitution that instituted a democratic system. However, the Prince of Monaco retains more direct political influence than most other constitutional monarchs. The city-state's population density is the largest in the world, and its life expectancy is Europe's highest.

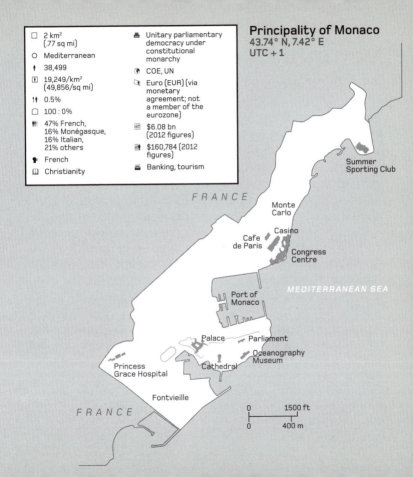

Principality of Monaco
43.74° N, 7.42° E
UTC + 1

- □ 2 km² (.77 sq mi)
- O Mediterranean
- † 38,499
- ⊞ 19,249/km² (49,856/sq mi)
- ↑↑ 0.5%
- ◯ 100 : 0%
- ♀♀ 47% French, 16% Monégasque, 16% Italian, 21% others
- ♟ French
- ▣ Christianity
- ♞ Unitary parliamentary democracy under constitutional monarchy
- ◉ COE, UN
- ⌥ Euro (EUR) (via monetary agreement; not a member of the eurozone)
- ▦ $6.08 bn (2012 figures)
- ▦ $160,784 (2012 figures)
- ▦ Banking, tourism

FRANCE

Summer
Sporting Club

Monte
Carlo

Casino

Cafe
de Paris

Congress
Centre

MEDITERRANEAN SEA

Port of
Monaco

Palace — Parliament

Oceanography
Museum

Princess
Grace Hospital

Cathedral

Fontvieille

FRANCE

| 0 | 1500 ft |
| 0 | 400 m |

Denmark

Geographically, Denmark covers most of the Jutland Peninsula, as well as an archipelago of over 400 islands to the east. A major centre of Viking activity, Denmark coalesced as a kingdom in the tenth century. In 1397, Sweden, Norway and Denmark united under Danish leadership, to form the Kalmar Union (named after the Swedish city in which the treaty of unification was signed). Sweden left this union in 1523, but Denmark and Norway continued until 1814. Denmark became a constitutional monarchy in 1849, establishing a parliamentary system of democratic government. The country's welfare system dates back to 1933 and, today, Denmark has the highest personal income tax in the world.

Since the Second World War, during which Denmark was invaded and occupied by Germany, the country has become one of the most socially advanced in the world, and a global leader in education, health care, civil liberties, social mobility and income equality. The Danish realm also includes the Faroe Islands and Greenland, which were granted home rule in 1948 and 1979, respectively.

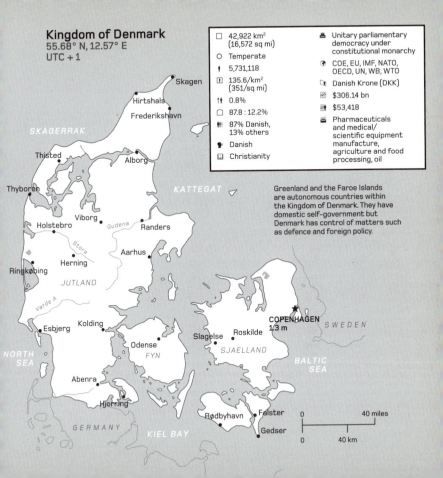

Kingdom of Denmark
55.68° N, 12.57° E
UTC + 1

☐	42,922 km² (16,572 sq mi)
O	Temperate
⚥	5,731,118
▯	135.6/km² (351/sq mi)
⚥	0.8%
◠	87.8 : 12.2%
⚥	87% Danish, 13% others
⚐	Danish
▢	Christianity

⚓	Unitary parliamentary democracy under constitutional monarchy
◉	COE, EU, IMF, NATO, OECD, UN, WB, WTO
▢	Danish Krone (DKK)
▤	$306.14 bn
▥	$53,418
⚒	Pharmaceuticals and medical/scientific equipment manufacture, agriculture and food processing, oil

Greenland and the Faroe Islands are autonomous countries within the Kingdom of Denmark. They have domestic self-government but Denmark has control of matters such as defence and foreign policy.

Skagen

Hirtshals
Frederikshavn

SKAGERRAK

Thisted

Thyborøn

Alborg

KATTEGAT

Viborg   Gudena
Holstebro   Randers

Stora

Ringkøbing   Aarhus

Herning

JUTLAND

Varde A

Esbjerg   Kolding

Odense

FYN

Abenra

Hjørring

GERMANY   KIEL BAY

NORTH SEA

COPENHAGEN
1.3 m   SWEDEN

Slagelse   Roskilde

SJAELLAND

BALTIC SEA

Rødbyhavn   Falster
Gedser

0		40 miles
0		40 km

Liechtenstein

A mountainous country bordered by Switzerland and Austria, Liechtenstein became a principality within the Holy Roman Empire in 1719. Its first prince, Anton Florian, was given the title by the emperor as a reward for his loyal service to the Habsburgs, although he chose to reside near Vienna. When the Holy Roman Empire was dissolved in 1806, Liechtenstein became an independent state, adopting a policy of neutrality in 1868. The country has been a constitutional monarchy since 1921, at which time democracy was established. It was not until 1938, following German annexation of Austria, that a ruling prince of Liechtenstein, Franz Joseph II, took up permanent residence. During the second half of the 20th century, the prince transformed Liechtenstein from an agricultural country into one of the world's wealthiest. This owed much to the establishment of specialized manufacturing firms and the country's status as a centre of international banking. In 2003, a new constitution gave the prince wider powers, including the ability to veto laws and dismiss governments.

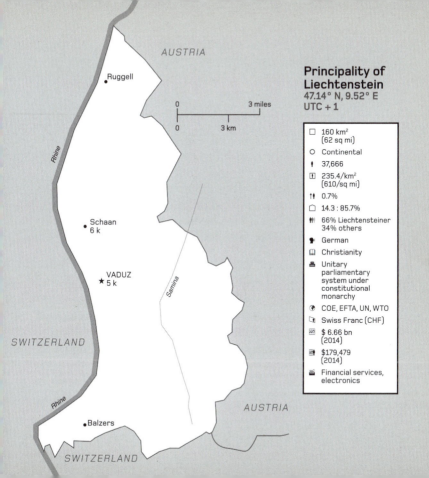

Principality of Liechtenstein
47.14° N, 9.52° E
UTC + 1

☐ 160 km²
(62 sq mi)

○ Continental

† 37,666

⊞ 235.4/km²
(610/sq mi)

†† 0.7%

◠ 14.3 : 85.7%

♯ 66% Liechtensteiner
34% others

♞ German

▢ Christianity

♜ Unitary
parliamentary
system under
constitutional
monarchy

◉ COE, EFTA, UN, WTO

⬚ Swiss Franc (CHF)

▦ $ 6.66 bn
(2014)

▦ $179,479
(2014)

⛴ Financial services,
electronics

Austria

From 1276, Austria was the power base of the House of Habsburg, a dynasty that ruled vast swathes of Europe. After 1521, the Habsburg lands were divided; a senior branch ruled Spain and its empire while a junior branch remained in Austria. The Austrian Habsburgs went on to build up a multinational realm that covered much of central and eastern Europe; Vienna became a cultural capital, famed for its spectacular architecture. In 1804, these Habsburg realms became the Austrian Empire, reorganized into the Austro-Hungarian Empire in 1867 – a dual monarchy that collapsed following defeat in the First World War. In 1919, the country was declared a federal republic. However, Germany annexed Austria in 1938 and, for a decade after the Second World War, the country was under Allied occupation. The Austrian State Treaty of 1955 restored national sovereignty, re-establishing the country's previous democratic system. On 26 October 1955, Austria declared perpetual neutrality – a date that is celebrated annually. Austria joined the EU in 1995, and has used the euro since its inception in 1999.

Republic of Austria
48.21° N, 16.37° E
UTC +1

- ☐ 83,879 km² (32,386 sq mi)
- ○ Temperate
- ♂ 8,747,358
- ⊞ 106/km² (275/sq mi)
- ♂♂ 1.3%
- ◻ 66.0 : 34.0%
- ♀♀ 91% Austrian, 9% others
- ☺ German

- 🏛 Christianity
- ♟ Federal parliamentary republic
- ⊕ EU, IMF, OECD, UN, WB, WTO
- 💶 Euro (EUR)
- 📊 $ 386.43 bn
- 💵 $44,177
- 🏭 Machinery and automotive manufacture

CZECH REPUBLIC

SLOVAKIA

GERMANY

Krems

Linz

Steyr

VIENNA ★
1.8 m

Salzach

Donau

Emns

Salzburg

Kufstein

Bregenz

Bischofshofen

Radstadt

St Anton

Inn

Innsbruck

A   L   P   S

Mur

Graz

Villach

Klagenfurt

Drau

HUNGARY

LIECHT.

SWITZERLAND

ITALY

SLOVENIA

0   100 miles

0   100 km

Sweden

The Swedish landscape is forested and mountainous, with a long stretch of coast to the east. The country is rich in mineral resources, most importantly copper. Established in the early 12th century, the Kingdom of Sweden united with Norway and Denmark in 1397 to form the Kalmar Union. Sweden left the union in 1523 and became a great regional power – at its mid-17th-century peak, its empire included all of Finland, as well as territory in the Baltics and northern Germany. By 1809, this empire had been lost, but a dynastic union saw Sweden control Norway for much of the 19th century. Since establishing a policy of neutrality in 1812, Sweden has not participated in armed conflict aside from a brief involvement in the Napoleonic Wars (1813–14); neither has it joined NATO. The monarch played a major role in governing the nation until 1917, when an elected left-wing government took away royal political power (this was not officially enshrined in law until 1974). Other major reforms followed, including equality of suffrage in 1919. Sweden became a member of the EU in 1995. Today, it has developed a wide-ranging social-welfare system that includes universal health care and free education.

Kingdom of Sweden
59.33° N, 18.07° E
UTC +1

□	447,420 km² (173,750 sq mi)
○	Temperate in south, Subarctic in north
♀	9,903,122
⊞	24.3/km² (63/sq mi)
♈	1.1%
▢	86.0 : 14.0%
♔	80% Swedish, 20% others
♟	Swedish
📖	Christianity
⚖	Unitary parliamentary system under constitutional monarchy
⊕	COE, EU, IMF, OECD, UN, WB, WTO
💱	Swedish Krona (SEK)
📈	$511.0 bn
📋	$51,560
🏭	Machinery and motor vehicles, telecommunications, paper manufacture

NORWEGIAN SEA

FINLAND

Kiruna • Tornealven

KJÖLEN MTS

Jokkmokk •

Luleälven

Boden •
Luleå

Skelleftealven

Storuman •

Umealven

Skellefteå •

Umea •

FINLAND

Ostersund •

Ljusnan

Sundsvall •

Hudiksvall •

GULF OF BOTHNIA

Mora •

Gavle •

Borlange •

Uppsala •

Västeras •
Karlstad • Örebro •
Vanern

STOCKHOLM
1.5 m

Norrköping •

Linköping •
Vattern

GULF OF FINLAND

Jonköping •

GOTLAND

Gothenburg •

SKAGERRAK

Varberg •

NORTH SEA

Halmstad •

DENMARK

Helsingborg •

Malmö •

BALTIC SEA

LATVIA

LITHUANIA

0 ___ 100 miles
0 ___ 100 km

NORWAY

Czech Republic

Originally the Kingdom of Bohemia, Czech territory was inherited by the Habsburg dynasty in 1526. This marked the beginning of nearly four centuries of Habsburg rule, which continued when their vast realm in eastern and central Europe was reorganized to form the Austro-Hungarian Empire in 1867. From the 19th century, Bohemia was an important industrial area and a major producer of coal. Following the collapse of Austria–Hungary in 1918, Czech and neighbouring Slovak territories joined to form the independent state of Czechoslovakia. From 1939 to 1945, Germany occupied the Czech part of the state, while Slovakia became independent. Reunited after the Second World War, Czechoslovakia then fell under Soviet influence with single-party communist rule. Despite a brief period of liberalization in 1968, known as the Prague Spring, communism remained in place until 1989, when a series of mass demonstrations – the Velvet Revolution – peacefully restored democratic rule. Czechoslovakia was dissolved in 1993, forming the Czech Republic and Slovakia. The Czech Republic joined NATO in 1999 and the EU in 2004.

☐	78,870 km² (30,452 sq mi)	🕮	Christianity
○	Temperate	🏛	Unitary parliamentary constitutional republic
�りり	10,561,633		
🈵	136.8/km² (355/sq mi)	🌐	COE, EU, IMF, NATO, OECD, UN, WB, WTO
♀♂	0.1%	💱	Czech Koruna (CZK)
◯	73.0 : 27.0%	💵	$192.92 bn
🙀	92% Czech, 8% others	💰	$18,267
🐾	Czech	🚗	Automobile manufacture

Czech Republic (or Czechia)
50.08° N, 14.44° E
UTC +1

San Marino

Entirely surrounded by Italy, San Marino is located on the slopes of Mount Titano in the Apennines. The country takes its name from St Marinus, a stonemason who founded a monastery on the site of the country in 301 CE. The territory became a republican city-state and retained its independence even after Italian unification in the 19th century. Originally, San Marino was ruled by the Arengo, an assembly of notable families, but this changed in 1243 with the establishment of a citizen-elected assembly. The current constitution is largely based on a series of statutes that came into force in 1600 – the world's oldest constitutional documents still in place. These provide for a parliament, called the Grand and General Council, which is elected every five years. Twice a year, parliament chooses two 'captains regent' to act as dual heads of state. San Marino was neutral during both world wars and has no regular armed forces; national defence is left to Italy. Tourism is vital to the Sammarinese economy, and its stamps and coins are highly valued by collectors.

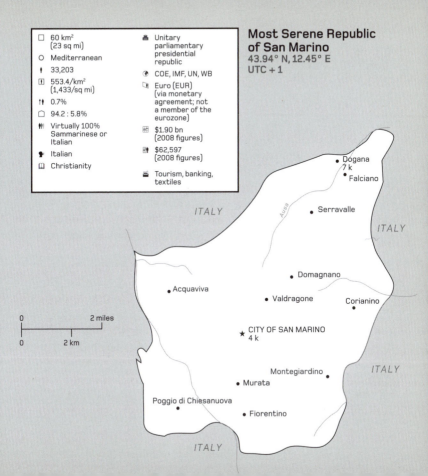

Most Serene Republic of San Marino

43.94° N, 12.45° E
UTC + 1

- ☐ 60 km² (23 sq mi)
- ○ Mediterranean
- ✝ 33,203
- ▦ 553.4/km² (1,433/sq mi)
- ⇅ 0.7%
- ◻ 94.2 : 5.8%
- ⚥ Virtually 100% Sammarinese or Italian
- ♟ Italian
- ◫ Christianity

- ⚓ Unitary parliamentary presidential republic
- ⊕ COE, IMF, UN, WB
- ▱ Euro (EUR) (via monetary agreement; not a member of the eurozone)
- ▦ $1.90 bn (2008 figures)
- ▦ $62,597 (2008 figures)
- ⛴ Tourism, banking, textiles

ITALY

ITALY

ITALY

ITALY

Ausa

• Dogana
  7 k
  • Falciano

• Serravalle

• Domagnano

• Acquaviva

• Valdragone

Corianino

★ CITY OF SAN MARINO
  4 k

Montegiardino •

• Murata

Poggio di Chiesanuova •

• Fiorentino

0 ——————— 2 miles

0 ——————— 2 km

Vatican City

The smallest sovereign state in the world, Vatican City is an enclave in Rome. It has only five entrances, which are guarded by the state's hundred-strong security force, the Pontifical Swiss Guard. The head of state is the Pope, elected by the College of Cardinals on the death (or resignation) of his predecessor. The Pope acts as an absolute monarch over the state, as well as the leader of the Catholic Church worldwide.

Vatican City is located on the burial site of the first pope, St Peter, who died in the mid-first century. From the mid-eighth century, the pope also ruled the Papal States, which at their peak in the 16th century covered much of central Italy. Over time the size of the Papal States diminished until, in 1870, Italy annexed Rome itself. Disagreements over Vatican City's status were not settled until 1929, when the Holy See and the Italian government signed the Lateran Treaty. This recognized Vatican City as an independent state and established its current territory.

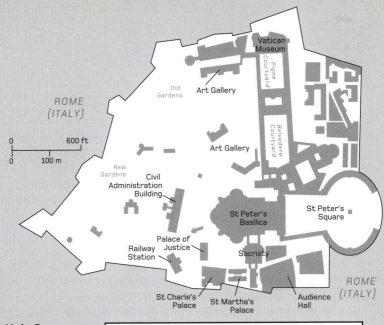

ROME
(ITALY)

Vatican Museum

Pigne Courtyard

Art Gallery

Old Gardens

0 | 600 ft
0 | 100 m

Belvedere Courtyard

Art Gallery

New Gardens

Civil Administration Building

St Peter's Basilica

St Peter's Square

Palace of Justice

Railway Station

Sacristy

ROME
(ITALY)

St Charle's Palace

St Martha's Palace

Audience Hall

**The Holy See
(Vatican City State)**
41.90° N, 12.45° E
UTC + 1

☐ 0.44 km² (0.17 sq mi)	⚥ Main groups are Italian and Swiss, but composition is multi-ethnic	◉ UN (permanent observer status)	
○ Temperate		⌨ Euro (EUR)	
♦ 800	♟ Italian, Latin	▦ $308.00 m (2011)	
⊞ 2,272/km² (5,885/sq mi)	⌂ Christianity	▦ $365,796 (2011)	
↿↾ 1.1%	♟ Ecclesiastical elective monarchy	⚱ Tourism, donations	
☐ 100 : 0%			

Slovenia

Nearly two-thirds of Slovenia's mountainous landscape is forested, intersected by rivers and lakes. In the 14th century, the region fell under the control of the Habsburgs, who ruled until the collapse of the Austro-Hungarian empire in 1918. Slovenian territory then became part of the new Kingdom of Yugoslavia – a merging of Serbia with the Slovene, Croat and Serb territories that had previously been part of Austria–Hungary. During the Second World War, Germany invaded and divided the territory between its neighbours. From 1945, Slovenia was a member of the Socialist Federal Republic of Yugoslavia, a single-party state comprising six republics formed after the war. In 1991, multi-party elections were followed by a ten-day war that saw Slovenia break away from Yugoslavia, escaping much of the bloodshed that occurred as the rest of the federation broke up. The Slovenian economy transitioned successfully from socialism to capitalism, and is one of the strongest in the region. A stable democracy today, Slovenia was the first former Yugoslav republic to join the EU, in 1994, and adopted the euro five years later.

☐	20,270 km² (7,826 sq mi)	📖	Christianity
○	Ranges from Mediterranean to continental	⚖	Unitary parliamentary constitutional republic
✝	2,064,845	◉	COE, EU, IMF, NATO, OECD, UN, WB, WTO
⊞	102.5/km² (266/sq mi)	💶	Euro (EUR)
⇈	0.1%		$43.99 bn
⬭	49.6 : 50.4%		$21,305
⇈⇈	83% Slovene 17% others	🚆	Machinery, transportation, textiles
♟	Slovene		

Republic of Slovenia
46.06° N, 14.51° E
UTC +1

Croatia

Crescent-shaped Croatia comprises flat plains to the north and an Adriatic coastline dotted with more than 1,100 islands and islets to the south. The Croats descend from a Slavic ethnic group that originated in Eastern Europe and migrated south in the sixth and seventh centuries, to settle the Balkan region of southeastern Europe. The Kingdom of Croatia was established in 925 CE, and united with Hungary in 1102. Under Habsburg rule from 1527, the country later became part of the Austro-Hungarian empire. In 1918, Croatia merged into the newly created Kingdom of Yugoslavia (literally 'South Slavia'); it functioned as a German puppet state from 1941–45, but adopted socialism following the Second World War, becoming one of the six Yugoslav republics. In 1991, Croatia declared independence, but it took four years of warfare for this to become established. During this period, most ethnic Serbs left the country, leaving a predominantly Croat population. The state initially had a semi-presidential structure of government but, since 2000, has adopted a parliamentary system. Following a ten-year application process, Croatia joined the EU in 2013.

Republic of Croatia
45.82° N, 15.98° E
UTC +1

□	56,590 km² (21,850 sq mi)	🐾	Croatian
○	Ranges from Mediterranean to continental	📖	Christianity
✝	4,170,600	🏛	Unitary parliamentary constitutional republic
田	74.5/km² (193/sq mi)	⌖	COE, EU, IMF, NATO, UN, WB, WTO
↕↑	-0.8%	💱	Croatian Kuna (HRK)
▱	59.3 : 40.7%	📊	$50.43 bn
👥	90% Croat 10% others	📈	$12,091
		🚂	Transportation equipment

Poland

At the centre of continental Europe, Poland is bounded by the Baltic Sea to the north and the Carpathian Mountains to the south. The first Polish state was founded in 966 CE and became part of the Polish–Lithuanian Commonwealth in 1569. In the late 18th century, Polish territory was divided between Austria, Prussia and Austria, wiping the country from the map. It re-emerged as an independent state in 1918, but was invaded and occupied by Germany and the USSR during the Second World War; six million Poles were killed, around half of them Jewish. From 1945, Poland became a satellite state of the Soviet Union, though it was more progressive than some others. 'Solidarity', an independent trade union, became a national political force during the 1980s, with nearly ten million members. Through nonviolent protest, the movement forced the regime to hold elections in 1989 and 1990. A Solidarity-led coalition won power and ushered in multiparty democratic rule, replacing the socialist economy with a capitalist system. The country joined the EU in 2004, and is currently one of the largest and most dynamic economies in Central Europe.

Republic of Poland
52.23° N, 21.01° E
UTC +1

- ☐ 312,680 km² (120,726 sq mi)
- ○ Temperate
- ♦ 37,948,016
- ▯ 123.9/km² (321/sq mi)
- ⥮ -0.1%
- ⌂ 60.5 : 39.5%
- ⋔ 97% Polish 3% others
- ☗ Polish
- ▥ Christianity

- ⚒ Unitary semi-presidential republic
- ⊛ COE, EU, IMF, NATO, OECD, UN, WB, WTO
- ▱ Polish Zloty (PLN)
- ▨ $469.51 bn
- ▤ $12,372
- ⚒ Machinery and transport equipment, iron and steel, mining (particularly coal)

Malta

The Republic of Malta comprises an archipelago of three main islands. The country is named after the largest, Malta, with two smaller inhabited islands (Comino and Gozo) to the northwest. Owing to its strategic position in the centre of the Mediterranean Sea, Malta has long been coveted as a naval base.

The first people to colonize Malta were Phoenician traders around 1000 BCE. Over subsequent centuries, the island was conquered by a series of powers, including the Carthaginians, Greeks, Romans, Arabs, Normans, Spanish and French. In 1814, the islands became part of the British empire, and a vital point on the maritime trade route to Asia. From 1940–42, during the Second World War, Malta was the subject of sustained bombing from German and Italian forces. The courage and resilience of the Maltese people led to the entire country being awarded Britain's George Cross for valour; an image of the medal features on Malta's national flag. The country gained independence in 1964 and was admitted to the EU in 2004.

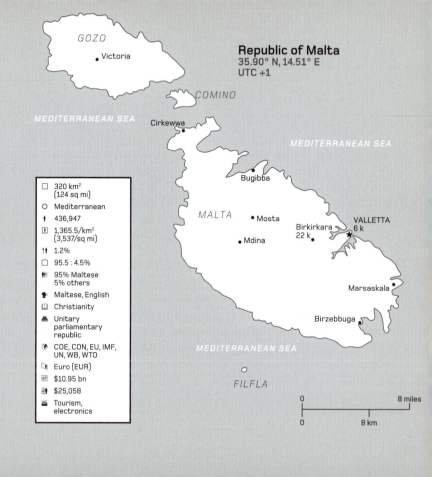

GOZO
• Victoria

COMINO

Republic of Malta
35.90° N, 14.51° E
UTC +1

MEDITERRANEAN SEA

Cirkewwa

MEDITERRANEAN SEA

Bugibba

MALTA

• Mosta

Birkirkara
22 k

VALLETTA
6 k

• Mdina

Marsaskala

Birzebbuga

MEDITERRANEAN SEA

FILFLA

- □ 320 km² (124 sq mi)
- ○ Mediterranean
- ♦ 436,947
- ▥ 1,365.5/km² (3,537/sq mi)
- ↑↑ 1.2%
- ◠ 95.5 : 4.5%
- ♦♦♦ 95% Maltese 5% others
- ♦ Maltese, English
- ▥ Christianity
- ♣ Unitary parliamentary republic
- ✪ COE, CON, EU, IMF, UN, WB, WTO
- ▱ Euro (EUR)
- ▦ $10.95 bn
- ▨ $25,058
- ▬ Tourism, electronics

0   8 miles
0   8 km

Bosnia and Herzegovina

L ocated in the western Balkans, Bosnia and Herzegovina is mountainous with many forested areas and a 19-kilometre (12-mile) coastline on the Adriatic Sea. Settled by various cultures such as the Illyrians and Celts in ancient times, the region was contested by rival powers, including local Slavic dynasties, until it was conquered by the Muslim Ottoman Empire in the mid-15th century. Under their rule, many people converted to Islam. Annexed by the Austro-Hungarian Empire in 1878, the territory later became part of the Kingdom of Yugoslavia in 1918, and one of six Yugoslav republics after the Second World War. In 1992, a declaration of independence triggered three years of civil war between Eastern Orthodox Serbs, Muslim Bosniaks and Catholic Croats – one of the most savage conflicts of the 20th century, with widespread ethnic cleansing and violence against civilians. The 1995 Dayton Accords ended the fighting by establishing two political entities within the country: the Bosniak–Croat Federation of Bosnia and Herzegovina (made up of ten cantons) and the Bosnian Serb Republic. The two regions are roughly the same size and mostly autonomous, with their own governments and parliaments.

Bosnia and Herzegovina
43.86° N, 18.41° E
UTC + 1

▢	51,210 km² (19,772 sq mi)	🕮	Islam
◯	Ranges from Mediterranean to continental	⚖	Federal parliamentary republic
✝	3,516,816	◉	COE, G77, IMF, UN, WB, WTO (observer)
⊞	60.1/km² (156/sq mi)	💱	Bosnia and Herzegovina Convertible Mark (BAM)
↕	-0.5%	💰	$16.56 bn
◠	39.9 : 60.1%	📄	$4,709
⚥	50% Bosniak 31% Serb 15% Croat 4% others	⛏	Metals
☻	Bosnian, Croatian, Serbian		

Slovakia

A landlocked Central European nation, northern Slovakia forms part of the Carpathian Mountains, while the south is mainly lowland. Slavic tribes settled the area in the fifth century CE; it became part of the Kingdom of Hungary in the tenth century and of the Austro-Hungarian Empire in 1867. Despite centuries of foreign rule, Slovakia has maintained a distinct cultural identity. Although it became part of Czechoslovakia in 1918, the country functioned as the independent Slovak Republic from 1939–45, acting as a Nazi German client state. Reunited as Czechoslovakia after the Second World War, it became part of the Eastern bloc and was subject to communism. The communist regime was ejected in 1989, to be replaced with a democratic government; four years later Czechoslovakia peacefully separated in a process known as the Velvet Divorce. The early years of independence were troubled by corruption, slow economic growth and increasingly authoritarian government. Elections held in 1998 saw a new government take power, followed by a series of political and economic reforms. These paved the way to creating a stronger economy, and facilitated Slovakia's entry to the EU in 2004.

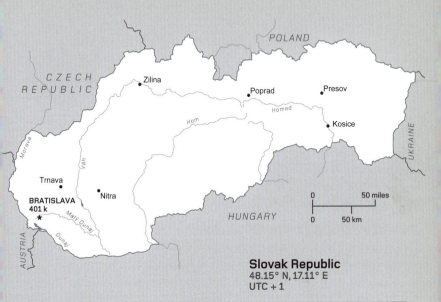

Slovak Republic
48.15° N, 17.11° E
UTC + 1

☐	49,035 km² (18,933 sq mi)	♛	81% Slovak 19% others	▣	Euro (EUR)
○	Temperate	☗	Slovak	▨	$89.55 bn
⊞	5,428,704	▢	Christianity	▨	$16,496
⊡	112.9/km² (293/sq mi)	♞	Parliamentary republic	▰	Metals, machinery and electrical equipment
⚲	0.1%	◉	COE, EU, IMF, NATO, OECD, UN, WB, WTO		
◠	53.5 : 46.5%				

Hungary

Mostly comprising plains, the Hungarian landscape is traversed by two major rivers: the Danube and the Tisza. The Magyars, a nomadic people who had migrated west from the Urals, conquered the territory in the late ninth century. After 1526, it fell under Habsburg rule and, from 1867, it was one of the main components of the Austro-Hungarian Empire. Hungary became independent after the First World War, albeit with territorial and population losses.

Following the Second World War, the country became a socialist republic – Soviet military intervention crushed a national revolt against communist rule in 1956. The government began political and economic reforms in 1968 – applying a blend of one-party socialism with market liberalization and decreased authoritarianism, known as 'Goulash Communism'. The country became a democratic republic in 1989, with free elections held the following year. It has alternately elected left- and right-wing leaders, who have focused on integrating with western Europe, regardless of political affiliation. Hungary joined the EU in 2004.

Hungary
47.50° N, 19.04° E
UTC + 1

☐ 93,030 km² (35,919 sq mi)	🏛 Hungarian	💷 Hungarian Forint (HUF)
○ Temperate	🏛 Christianity	💵 $124.34 bn
👤 9,817,958	🏛 Unitary parliamentary constitutional republic	💵 $12,665
🏢 108.5/km² (281/sq mi)		🏭 Mining, metals, machinery and equipment manufacture
↕↕ -0.3%	⚙ COE, EU, IMF, NATO, OECD, UN, WB, WTO	
☐ 71.7 : 28.3%		
👥 86% Hungarian 14% others		

Montenegro

Despite its small size, Montenegro has a varied landscape that descends from rugged mountains in the north to a coastal plain along the Adriatic Sea. During the Middle Ages, Montenegro was dominated by the Byzantines and the Serbians, while the Venetians ruled coastal areas from 1420 to 1797 – the country's name is adapted from the Italian for 'black mountain', and describes Mount Lovćen, to the southwest of the country. From the late 15th century, the Ottoman empire tried to extend into the country but failed to defeat its mountain clans. Much of Montenegro remained independent as an ecclesiastical principality ruled by an Eastern Orthodox bishop until the late 19th century, when it became a unified secular principality, fighting a series of wars that led to independence. In 1918, Montenegro became part of Yugoslavia and was one of its constituent parts when it became a socialist federal republic in 1945. When Yugoslavia broke up in 1992, Montenegro remained federated with Serbia until a 2006 referendum saw a 55 per cent majority vote for leaving this union – Montenegro declared its independence and became a sovereign state.

BOSNIA-HERZEGOVINA

SERBIA

• Pljevlja

Tara

Bijelo Polje •

Piva

Kolasin •   Berane •

• Niksic

KOSOVO
(SERBIA)

CROATIA

Herceg Novi •

PODGORICA
165 k
★

• Kotor

ALBANIA

Lake
Shkoder

ADRIATIC
SEA

• Bar

Montenegro
42.43° N, 19.26° E
UTC +1

| 0 | | 30 miles |
| 0 | | 30 km |

☐ 13,810 km²
     (5,332 sq mi)

○ Mediterranean

† 622,781

⊞ 46.3/km²
     (120/sq mi)

†† 0.1%

☐ 64.2 : 35.8%

†† 45% Montenegrin
      29% Serbian
      26% others

🖌 Montenegrin

▥ Christianity

♗ Unitary parliamentary
     republic

⊕ COE, IMF, NATO, UN,
     WB, WTO

▢ Euro (EUR)
     (adopted
     unilaterally; not a
     formal member of
     the eurozone)

▦ $ 4.17 bn

▨ $6,701

▤ Metals

Serbia

The largest country in the Balkans, Serbia has fertile plains to the north, a hilly central terrain traversed by rivers, and mountains to the south. A major regional power during the 14th century, the Serbian empire was conquered by the Ottomans in the mid-15th century. The Ottomans themselves ruled until the early 19th century, when Serb rebels overthrew them, gaining independence as a kingdom. Serbia was one of several South Slavic states that joined to form the Kingdom of Yugoslavia in 1918, which the Axis powers invaded and partitioned during the Second World War. When Yugoslavia reunited as a one-party socialist federation in 1945, Serbia was its largest and most populous republic. Despite being communist, Yugoslavia left the Eastern bloc in 1948 to pursue an independent foreign policy. When the country split apart in 1991, Serbia and Montenegro stayed united as a federal republic, from which Montenegro seceded in 2006. In 2008, the southern Serbian region of Kosovo, mainly populated by ethnic Albanian Muslims, declared independence, but this has not been universally recognized, and Serbia does not accept it as a state.

Republic of Serbia
44.79° N, 20.45° E
UTC +1

- ☐ 88,360 km² (34,116 sq mi)
- ○ Ranges from Mediterranean to continental
- 7,057,412
- 80.7/km² (209/sq mi)
- -0.5%
- 55.7 : 44.3%
- 83% Serb 17% others
- Serbian
- Christianity
- Unitary parliamentary constitutional republic
- COE, IMF, UN, WB, WTO (observer)
- Serbian Dinar (RSD)
- $37.75 bn
- $5,348
- Motor vehicles, electrical machines, agriculture and food processing

HUNGARY

Subotica

DANUBE

CROATIA

Zrenjanin

Novi Sad

ROMANIA

BELGRADE
★ 1.2 m

Smederevo

BOSNIA-HERZEGOVINA

Kragujevac

Bor

Uzice

Nis

Novi Pazar

Leskovac

MONTENEGRO

Pristina

KOSOVO

BULGARIA

ALBANIA

MACEDONIA

The Kosovo region unilaterally declared its independence in 2008. Its independence is recognised by 113 UN member states, but not by Serbia.

0 ___ 30 miles

0 ___ 30 km

Albania

The Balkan territory of Albania is split between an interior dominated by rugged mountains, and lowlands along the country's Ionian and Adriatic coasts. In 1479, the Ottoman Empire gained control of the region, converting many to Islam. Facing national unrest, the Ottoman empire fell into decline from the late 19th century, and the Albanians declared independence in 1912. Autonomy was short-lived. In 1939, Italy invaded and annexed Albania; German occupation followed in 1943, although communist partisans had liberated the country by the end of 1944. Albania became a single-party communist state ruled by authoritarian and isolationist Enver Hoxha. In 1967, he declared Albania to be an atheist state, ordering the closure of mosques and churches. Following Hoxha's death in 1985, his successor Ramiz Alia liberalized Albania, holding multiparty elections in 1991, which ended communist rule. In 1996, the collapse of a pyramid scheme in which one-quarter of the population had invested threw Albania into economic and political crisis. The country has since recovered and, in 2014, became a candidate for EU membership.

Republic of Albania
41.33° N, 19.82° E
UTC + 1

MONTENEGRO

KOSOVO
(SERBIA)

Drin

• Shkodër   Kukës •

MACEDONIA

ADRIATIC SEA

★ TIRANA
454 k

• Durrës

Elbasan •

Devoll

Lake Ohrid

Lake Prespa

Berat •

Osum

Korcë •

Vlorë •

Vjose

Gjirokastër •

GREECE

0   50 miles

0   50 km

GREECE

▢	28,750 km² (11,100 sq mi)
O	Mild temperate
♀	2,876,101
⊞	105/km² (279/sq mi)
↑↑	-0.2%
▢	58.4 : 41.6%
♟	97% Albanian 3% others
☻	Albanian
▣	Islam
⚒	Unitary parliamentary constitutional republic
◉	COE, IMF, NATO, UN, WB, WTO
◱	Albanian Lek (ALL)
▦	$ 11.93 bn
♙	$4,147
⛟	Agriculture

Russia

- 17,098,250 km² (6,601,671 sq mi)
- Ranges from steppe in south to humid continental in west to subarctic in Siberia
- 144,342,396
- 8.8/km² (23/sq mi)
- 0.2%
- 74.1 : 25.9%
- 78% Russian 22% others
- Russian

- Christianity
- Federal semi-presidential constitutional republic
- ACD, CIS, COE, EAEU, IMF, SCO, UN, WB, WTO
- Russian Ruble (RUB)
- $ 1.28 tn
- $8,748
- Oil and gas, mining, chemicals and metals manufacture

NORWAY
ARCTIC OCEAN
Severnaya Zemlya
New Siberian Islands
LAPTEV SEA
Murmansk
BARENTS SEA
KARA SEA
Novaya Zemlya
Tiksi
Dikson

BALTIC SEA
FINLAND
Kaliningrad
POLAND
EST
St Petersburg
Arkhangel'sk
Pechora
Salekhard
Igarka
Vilyuysk
Yakutsk
Lena

LITH
LAT
Novgorod
Smolensk
Kotlas
Pechora
U
R
A
L
S

BELARUS
MOSCOW
12.2 m
Kirov'
Perm
Surgut
Lensk
Aldan
Neryungr

UKRAINE
Orel
Nizhniy-Novgorod
Kazan
Yekaterinburg
Lesosibirsk
Angara
Bratsk

Kursk
Voronezh
Ufa
Chelyabinsk
Irtysh
Tomsk
Kansk
Chita
Lake Baikal

Don
Saratov
Samara
Kurgan
Yenisey
Krasnoyarsk
Ulan-Borzya

Rostov
Volga
Orenburg
Omsk
Novosibirsk
Irkutsk
Ulan-Ude

Volgograd
Astrakhan'
KAZAKHSTAN
A
L
T
A
I
MONGOLIA

BLACK SEA
GEO
CASPIAN SEA
AZER

0   750 miles

0   750 km

CHINA

Russian Federation
55.75° N, 37.62° E
UTC + 2 to +12

The world's largest country, Russia stretches from Eastern Europe across Siberia to the Pacific, with territory ranging from Arctic desert and tundra to forests and steppe plains. Originating in the 14th-century expansion of the Grand Duchy of Moscow, Russia extended power and influence over a vast area; its rulers assumed the title of 'tsar' in 1547, and emperor in 1721. Their regime collapsed in the Russian Revolution of 1917 and communists seized power, establishing single-party rule. By 1922, Russia was the most influential component of the Union of Soviet Socialist Republics (USSR) which, in the wake of the Second World War, became one of two global superpowers and leader of the Eastern Bloc. In 1991, facing economic decline and independence movements, the USSR dissolved into 15 republics. The rapid attempted transition to democracy and free-market capitalism saw the country suffer rampant inflation and a prolonged recession in the 1990s. More recently, as the Russian economy has recovered (largely thanks to its rich deposits of oil and natural gas), its political leadership has grown more authoritarian.

Greece

The Greek mainland and its two peninsulas are rugged and mountainous, while one-fifth of its territory comprises some 2,000 islands. The classical Greek culture of the fifth and fourth centuries BCE was a major influence on Western arts, sciences, philosophy and political thought. From the early fourth century CE, Greece became part of the Byzantine Empire before falling under Ottoman rule in 1453. It regained independence in 1830, and over subsequent decades expanded its borders north and extended its rule over islands in the Aegean, Ionian and Mediterranean seas, the largest of which is Crete. In 1940, Greek armed forces repelled an Italian invasion, but German intervention led to the Axis powers occupying the country. After the Second World War, Greece descended into a civil war between communist and government forces; it took three years of fighting for the government to gain victory. Greece joined the EEC in 1981 and the eurozone in 2001; it has been in financial crisis since 2009, forcing the government to cut spending and increase taxes, as well as take out loans that have left the country heavily in debt.

Hellenic Republic
37.98° N, 23.73° E
UTC + 2

- ☐ 131,960 km² (50,950 sq mi)
- ○ Mostly temperate
- 👤 10,746,740
- ⊞ 83.4/km² (216/sq mi)
- 👥 -0.7%
- ☐ 78.3 : 21.7%
- 👪 88% Greek 12% others
- 👤 Greek

- 📖 Christianity
- ♟ Unitary parliamentary republic
- 🌐 COE, EU, IMF, NATO, OECD, UN, WB, WTO
- 💶 Euro (EUR)
- 💰 $ 194.56 bn
- 💵 $18,104
- 🚢 Shipping, tourism, agriculture

0   100 miles
0   100 km

Romania

Romania is formed of mountains, hills and plains in almost equal parts; the Carpathian Mountains run in an arc across its centre. The Kingdom of Romania, comprising the principalities of Wallachia and Moldavia, proclaimed independence from the Ottoman Empire in 1877. Supporting the Allies during the First World War, the country made territorial gains thereafter. Initially siding with Germany in the Second World War, Romania switched allegiance to the Allies in 1944. In 1947, the country became part of the Eastern Bloc, following the establishment of a communist people's republic. The USSR's militarily occupation (1944–58) drained many of Romania's oil, coal and uranium resources; thereafter, from 1965 to 1989, the state fell under Nicolae Ceaușescu's dictatorial and authoritarian rule. Ceaușescu was overthrown and executed in a popular revolution that brought communist rule to an end. Romania adopted a presidential system, and its first free elections since 1937 were held in 1990. Since that time, there has been a transition into a functioning multi-party democracy. Romania joined the EU in 2007, integrating it more fully into the western European economic and political system.

Romania
44.43° N, 26.10° E
UTC + 2

☐ 238,390 km² (92,043 sq mi)	♟ 83% Romanian 17% others	🗔 Romanian Leu (RON)
○ Temperate	🏛 Romanian	🕮 $186.69 bn
⊡ 19,705,301	🏛 Christianity	🗈 $9,474
⊡ 85.6/km² (222/sq mi)	⚒ Unitary semi-presidential republic	⚡ Electrical machinery and equipment, textiles manufacture
⇅ -0.6%	⊙ COE, EU, IMF, NATO, UN, WB, WTO	
◻ 54.7 : 45.3%		

Macedonia

Much of the landlocked Balkan state of Macedonia, with the exception of the central Vardar River valley, is rugged and mountainous. The territory was ruled by the Ottoman Empire for over five centuries (c. 1400–1912) before, in the early 20th century, Macedonia became part of Serbia and then Yugoslavia. It was a constituent part of the Socialist Federal Republic of Yugoslavia following the Second World War, seceding peacefully from the federation in 1991, to become an independent parliamentary democracy. In 2001, an armed insurgency from Albanian separatists in the north of the country led the government to grant the Albanian minority (around one-quarter of the populace) greater local rights and cultural recognition. Macedonia has been a candidate for EU membership since 2005, but is yet to join. An ongoing barrier has been Greek government objections to the country's name (shared with a region in northern Greece). After many years of using the provisional name of the 'Former Yugoslav Republic of Macedonia', a potential solution emerged in 2018, with both countries agreeing to propose a new name: the Republic of North Macedonia.

Republic of Macedonia*
42.00° N, 21.43° E. UTC + 1

☐ 25,710 km² (9,927 sq mi)	⬠ 57.2 : 42.8%	◉ COE, IMF, UN, WB, WTO
○ Mild continental	♛ 64% Macedonian, 25% Albanian 11% others	Macedonian Denar (MKD)
✝ 2,081,206		$ 10.90 bn
⊞ 82.5/km² (214/sq mi)	♟ Macedonian	$5,237
⇅ 0.1%	▯ Christianity	Agriculture and food processing, textile production
	⚒ Parliamentary republic	

* A proposed new name, the Republic of North Macedonia, awaits ratification.

Finland

Finland was part of Sweden from the late 13th century until 1809, when it became an autonomous grand duchy within the Russian empire. The country gained independence in 1917 and became a democratic republic. During the Second World War, it fought off a Soviet invasion and, despite losing territory to the USSR after the war, retained its sovereignty. Nearly three-quarters of Finland is forested, and it contains more than 60,000 lakes. Largely agricultural for much of its history, the country has experienced lower living standards than elsewhere in Europe. The economy industrialized from the 1950s and Finland joined the EU in 1995, becoming a founding member of the eurozone four years later. Although forestry and mining are still important, the country has become a world leader in the telecommunications and electronics industries. As prosperity has grown, Finnish society has remained one of the world's most egalitarian. A major factor is a world-class education system that provides free schooling to the entire population – a recent study has determined that Finland is now the happiest nation.

Republic of Finland
60.17° N, 24.94° E
UTC + 2

☐	338,420 km² (130,665 sq mi)
○	Cold temperate
✝	5,495,096
⊞	18.1/km² (47/sq mi)
⇈	0.3%
◫	84.4 : 15.6%
⚥	89% Finnish 11% others
☙	Finnish Swedish Sami
▫	Christianity
♨	Unitary parliamentary republic
◈	COE, EU, IMF, OECD, UN, WB, WTO
▭	Euro (EUR)
▦	$236.79 bn
▦	$43,090
⛟	Engineering, telecommunications, electronics, metal mining, timber

Lithuania

The largest and most populous of the Baltic states, coastal Lithuania is covered by sand dunes and lagoons, while the interior is mostly low-lying plains, dotted with lakes and rivers. The Grand Duchy of Lithuania was founded in the early 13th century, and eventually extended over much of modern-day Belarus and Ukraine. In 1386, it formed a union with Poland that became a dual monarchy, the Polish–Lithuanian Commonwealth, in 1569. During the late 18th century, the commonwealth was partitioned among its neighbours, with most of Lithuania being absorbed into the Russian Empire. Functioning as an independent republic from 1918–40, Lithuania was subsequently annexed by the USSR. In 1990, it became the first part of the USSR to declare independence, although Russian forces remained until 1993. Today, the country is a democratic republic that has integrated, politically and economically, with western Europe. Lithuania joined NATO and the EU in 2004, adopting the euro a year later. While the economy has recovered from the 2008 financial crisis, one long-term problem is a rapid rate of population decline as younger people have emigrated to find work elsewhere in Europe.

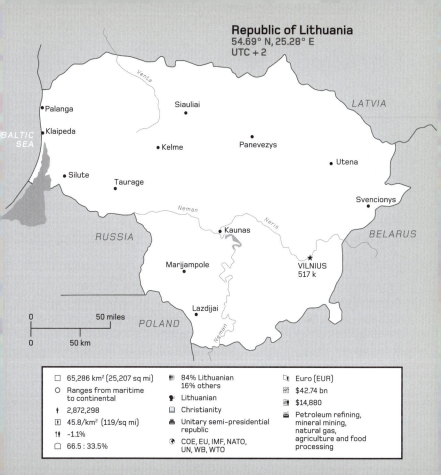

Republic of Lithuania
54.69° N, 25.28° E
UTC + 2

LATVIA

Venta

Palanga

Siauliai

Klaipeda

BALTIC
SEA

Kelme

Panevezys

Utena

Silute

Taurage

Neman

Neris

Svencionys

RUSSIA

Kaunas

BELARUS

Marijampole

VILNIUS
517 k

0	50 miles
0	50 km

Lazdijai

Neman

POLAND

☐	65,286 km² (25,207 sq mi)	⚥	84% Lithuanian 16% others	🏦	Euro (EUR)
O	Ranges from maritime to continental	👤	Lithuanian	📊	$42.74 bn
♰	2,872,298	⛪	Christianity	📈	$14,880
⊞	45.8/km² (119/sq mi)	🏛	Unitary semi-presidential republic	🏭	Petroleum refining, mineral mining, natural gas, agriculture and food processing
⇅	-1.1%	🌐	COE, EU, IMF, NATO, UN, WB, WTO		
☐	66.5 : 33.5%				

Latvia

The Latvian landscape, over half of which is forested, features highlands to the east and west that flank central lowlands. Owing to its position on the Baltic Sea and numerous navigable rivers, the country has played a vital role in European trade since the 13th century. Successively under German, Polish, Swedish and Russian control, Latvia finally declared independence after the First World War. In 1940, however, the USSR used military force to annexe Latvia and it became the Latvian Soviet Socialist Republic. An upsurge in nationalist sentiment in the late 1980s, and public demands for greater autonomy, culminated in a declaration of independence that was recognized by the USSR in 1991. Russia kept troops in Latvia until 1994, and more than a quarter of the country's population remains Russian-speaking. Democratic rule is now established, with parliamentary and presidential elections held every four years. Since independence, Latvia has looked increasingly westward, joining NATO and the EU in 2004 and the eurozone in 2014. The country's economy has transitioned to free-market capitalism, and it has become an important transportation hub.

Republic of Latvia
56.95 N 24.11 E. UTC + 2

☐ 64,490 km² (24,900 sq mi)	♔ 62% Latvian 26% Russian 12% others	◈ COE, EU, IMF, NATO, OECD, UN, WB, WTO	
○ Maritime	♙ Latvian	▤ Euro (EUR)	
♦ 1,960,424	▦ Christianity	▦ $27.68 bn	
▣ 34/km² (88.8/sq mi)	⚒ Unitary semi-presidential parliamentary constitutional republic	▦ $14,118	
↕ -0.4%		⚒ Transit services, forestry, agriculture, machinery and electronic manufacture	
☐ 67.4 : 32.6%			

Ukraine

With a landscape of wide, fertile plains, Ukraine was once known as the 'breadbasket of Europe' and remains a major exporter of grain today. It was the centre of power for the Kievan Rus, a federation of Slavic tribes that dominated the region during the 10th and 11th centuries. Most of its territory then became part of the Polish–Lithuanian Commonwealth. After more than 100 years of functional autonomy, a large area of Ukraine became part of the Russian Empire from 1764–95 (its remaining territory was absorbed by the Austro-Hungarian Empire). In 1917, the whole country won independence, only to be conquered by the USSR in 1920. Ukraine suffered two Soviet-engineered famines (1921–22 and 1932–33) and brutal German occupation during the Second World War.

Independent since 1991, the population remains polarized between those wanting to integrate with Europe and those seeking closer ties with Russia. Since 2014, there has been unrest in eastern Ukraine, between government and pro-Russian forces; during this period Russia annexed Crimea from Ukraine.

BELARUS

Chernihiv

RUSSIA

POLAND

Rivne

Zhytomyr

★ KIEV
2.9 m

Kharkiv

L'viv   Bug

Ternopil'

Bila Tserkva

Poltava

Donets

Lisichansk

Ivano-Frankivs'k

Vinnytsya

Cherkasy

Kremenchuk

Slovyans'k

Uzhhorod

Dniester

Southern Bug

Dnipropetrovs'k

Pavlograd

Luhans'k

Chernivtsi

Zaporizhzhya

Donets'k

MOLDOVA

Nikopol'

ROMANIA

Mariupol'

Mykolaïv

Melitopol'

0   100 miles

0   100 km

Odessa

Kherson

Dnieper

SEA OF AZOV

Izmayil

BLACK SEA

CRIMEA

Simferopol'

Sevastopol'

Russia effectively controls
Crimea following its 2014
annexation.

Ukraine
50.45° N, 30.52° E. UTC + 2

☐ 603,550 km² (233,032 sq mi)	78% Ukrainian 17% Russian 5% others	Ukrainian Hryvnia (UAH)
○ Temperate continental	Ukrainian	$93.27 bn
45,004,645	Christianity	$2,186
77.7/km² (201/sq mi)	Unitary semi-presidential constitutional republic	Mining (particularly iron ore and coal), agriculture (particularly grain), heavy industry
-0.3%	COE, CIS (associate), IMF, UN, WB, WTO	
69.9 : 30.1%		

Bulgaria

The Bulgarian landscape features alternate bands of fertile plain and mountains from north to south; in the east lies the sandy, subtropical Black Sea coast. The Bulgars originated in central Asia and migrated to the Balkans in the seventh century CE, joining the local Slavic population to create the first Bulgarian state. From 864, the country began to convert from paganism to Christianity, developing its own independent Bulgarian Orthodox church. Bulgaria was a regional power until 1396, when the Ottoman Empire took control. In 1878, northern territories achieved autonomy as the Principality of Bulgaria and, 30 years later, the entire country became an independent kingdom. Abolishing the monarchy in the wake of the Second World War, the country became a one-party communist state and the USSR's closest Balkan ally. Democracy returned in 1990, with multiparty elections leading to a new constitution a year later, and a democratic parliamentary government. Bulgaria abandoned socialist economic policies and moved towards a free-market system. At the same time it became more integrated with western Europe, joining NATO in 2004 and the EU in 2007.

Republic of Bulgaria
42.70° N, 23.32° E
UTC + 2

☐	111,000 km² (42,857 sq mi)	⚒	Unitary parliamentary constitutional republic
○	Temperate	☯	COE, EU, IMF, NATO, UN, WB, WTO
⚲	7,127,822		Bulgarian Lev (BGN)
⊞	65.7/km² (170/sq mi)		$52.40 bn
↑↓	-0.7%		$7,351
☐	74.3 : 25.7%	⚒	Electricity generation, mining (metal ore, coal, various minerals)
☵	77% Bulgarian 23% others		
☃	Bulgarian		
☐	Christianity		

Belarus

The Slavic country of Belarus was incorporated into the Grand Duchy of Lithuania from the 13th century and later became part of the Polish–Lithuanian Commonwealth. Annexed by Russia in 1795, it became the Byelorussian Soviet Socialist Republic following the formation of the USSR. The Second World War devastated the country, leaving cities in ruins and a population reduced by one-third. In 1991, the country changed its name to the Republic of Belarus, becoming an independent nation after the dissolution of the USSR later that year. Unlike other former Soviet states, Belarus did not undergo mass privatization after independence; 80 per cent of its industries remain state owned. It also retains closer ties to Russia than any other former Soviet republic; in 1999, the two countries signed a treaty to form a political union, but there has been little effort to integrate them since that time. Under the 1994 constitution, Belarus has a presidential system but only one man (Alexander Lukashenko) has held the office, having won five consecutive elections from 1994–2015. His government is highly authoritarian, with restrictions on freedom of speech and the press.

Republic of Belarus
53.90° N, 27.56° E
UTC + 3

☐ 207,600 km² (80,155 sq mi)	◻ 77.0 : 23.0%	🌐 CIS, EAEU, IMF, NAM, UN, WB, WTO (observer)
O Transitional between continental and maritime	⚎ 84% Belarusian 16% others	💷 Belarusian Ruble (BYN)
† 9,507,120	⚐ Belarussian	🔢 $47.43 bn
⬆ 46.9/km² (122/sq mi)	🏛 Christianity	💰 $4,989
↑↑ 0.2%	⚖ Unitary presidential republic	🏭 Machinery and equipment, automotive manufacture

Estonia

Estonia is geographically part of the Baltic region, although its people are Finnic in origin. Besides the mainland, it has around 1,500 islands and islets. The country was pagan until the early 13th century, when Christian armies conquered the territory. It was subsequently subjected to periods of German, Danish, Swedish and Russian rule, before finally becoming independent in 1918. After experiencing both Soviet and German occupation during the Second World War, in 1944 it was reoccupied by the Soviets, who annexed it to the USSR as the Estonian Soviet Socialist Republic.

The country reasserted its independence with the collapse of the USSR in 1991, but Russian forces remained in the country until 1994. Estonia has since become a stable, increasingly developed and prosperous, parliamentary democracy, joining the EU and NATO in 2004, and the eurozone in 2011. Estonians enjoy a high standard of living, with free education and universal health care. The country has also embraced digital technology; in 2005, it became the first country in the world to hold a general election online.

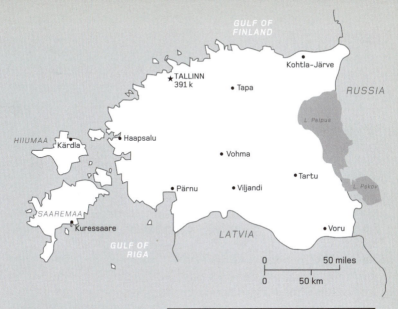

Republic of Estonia
59.44° N, 24.75° E
UTC + 2

☐ 45,230 km² (17,463 sq mi)	🏛 Christianity
○ Maritime	♟ Unitary parliamentary constitutional republic
✝ 1,316,481	
▯ 31.1/km² (81/sq mi)	🌐 COE, EU, IMF, NATO, OECD, UN, WB, WTO
†↑ 0.1%	💶 Euro (EUR)
☐ 67.5 : 32.5%	📊 $23.14 bn
▥ 69% Estonian 25% Russian 6% others	📈 $17,575
♟ Estonian	⚙ Electronics, engineering

Turkey

Straddling two continents, five per cent of Turkey's landmass lies within Europe and the rest in Asia. It was once the centre of the Ottoman Empire, which at its peak ruled much of central and southeastern Europe, North Africa and western Asia. Mustafa Kemal Atatürk founded the Republic of Turkey in 1923, embarking on a series of reforms that aimed to create a secular nation-state and moving the capital from Istanbul to Ankara. Turkey's first multiparty elections were held in 1946, and the first democratic transition of power from the ruling Republican People's Party occurred in 1950. The country has suffered periods of political instability, with military coups occurring in 1960, 1971 and 1980, although democratic rule was ultimately re-established after each one. Since 1978, the Turkish state has been involved in an ongoing conflict with Kurdish separatist forces in the southeast of the country. In recent years, many Turks have come to believe that the government has grown too authoritarian, and the military staged an unsuccessful coup attempt in 2016.

Republic of Turkey
39.93° N, 32.86° E
UTC + 3

☐ 785,356 km² (303,223 sq mi)	♟ Turkish	▨ $857.75 bn
○ Temperate	▱ Islam	▨ $10,788
✝ 79,512,426	♨ Unitary parliamentary constitutional republic	⚒ Textiles, automotive manufacturing petrochemical, electronics manufacture, agriculture
▣ 103.3/km² (268/sq mi)		
↑↑ 1.6%	☮ ACD, COE, IMF, NATO, OECD, UN, WB, WTO	
◻ 73.9 : 26.1%		
♟♟ 73% Turkish 19% Kurdish 8% others	▭ Turkish Lira (TRY)	

Moldova

Most Moldovan territory lies between two rivers: the Dniester to the northeast and the Prut, which forms the border with Romania to the west. The country has origins in the Bessarabia region of the Principality of Moldavia, a historic state that joined with Romania in 1859. Russia annexed Bessarabia in 1812 and it became independent in the aftermath of the Russian Revolution a century later. Bessarabia then unified with Romania, only to be ceded to the USSR in 1940, forming part of the Moldavian Soviet Socialist Republic. After the collapse of the Soviet Union in 1991, the country experienced a decade of financial depression as a newly independent state; despite improvements in economic performance, it remains one of Europe's poorest and least developed countries. Moldova has been further troubled by events in Transnistria, a breakaway region on the eastern border that saw military clashes in 1992. This region now has de facto independence, but is not recognized by any UN member. In 1998, Moldova became the only former Soviet state to elect a communist party with a ruling majority. Since 2009, however, it has been ruled by various pro-European coalitions.

Republic of Moldova
47.01° N, 28.86° E
UTC + 2

UKRAINE

Edinet

Nistru

Dniestr

Balti

UKRAINE

□ 33,850 km²
(13,070 sq mi)

○ Moderate
continental

♦ 3,552,000

⊞ 123.6/km²
(320/sq mi)

⚥ -0.1%

□ 45.1 : 54.9%

⚥ 75% Moldovan
25% others

☻ Romanian

▭ Christianity

▦ Unitary
parliamentary
constitutional
republic

◉ COE, CIS, IMF, UN,
WB, WTO

▯ Moldovan Leu (MDL)

▨ $6.75 bn

▨ $1,900

▤ Agriculture

Prut

Bîc

Nistru

CHIŞINĂU
725 k

★

Tiraspol

ROMANIA

Comrat

UKRAINE

Cahul

0   50 miles
0   50 km

Cyprus

An island in the eastern Mediterranean, Cyprus is ringed by coastal lowlands and has two main mountain ranges (the Troodos and the Kyrenia) separated by a central plain. Owing to its strategic position between Europe and Asia, it has been annexed by an array of imperial powers, including Assyria, Egypt, Persia, Macedonia and Rome. The Ottomans ruled from 1571, leading to an influx of Turkish settlers. In 1878, Cyprus fell under British rule for nearly a century; a decade of nationalist violence led to independence in 1960. Tension between Greek and Turkish Cypriots led to more violence from 1963 to 1974, when a Greek nationalist attempt to join Cyprus with Greece triggered a Turkish invasion and the subsequent occupation of the northeastern third of the island. This territory, the Turkish Republic of Northern Cyprus, declared independence in 1983. Only recognized by Turkey, the region has de facto independence, with a UN peacekeeping force patroling a buffer zone that divides the island. Since joining the EU in 2004 (and the eurozone in 2008), Cyprus has become increasingly prosperous, largely thanks to tourism.

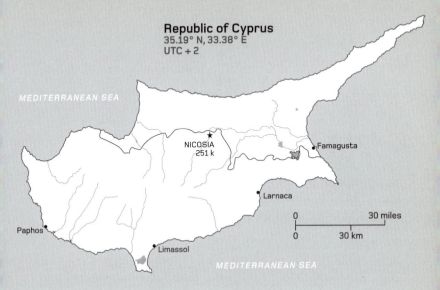

Republic of Cyprus
35.19° N, 33.38° E
UTC + 2

MEDITERRANEAN SEA

★ NICOSIA
251 k

Famagusta

Larnaca

Paphos

Limassol

MEDITERRANEAN SEA

0   30 miles

0   30 km

☐ 9,250 km² (3,571 sq mi)	♟ 74% Greek 25% Turkish 1% others	⊕ COE, CON, EU, IMF, UN, WB, WTO
○ Mediterranean	⚐ Greek, Turkish	⌨ Euro (EUR)
✆ 1,170,125	☐ Christianity, Islam	▦ $19.80 bn
⊞ 126.6/km² (328/sq mi)	⚒ Unitary presidential constitutional republic	▨ $23,324
⚦ 0.8%		Tourism, financial services, shipping
☐ 66.8 : 33.2%		

Africa

The birthplace of the human species 200,000 years ago, Africa comprises one-fifth of the world's landmass and contains one-seventh of its population. The climate is incredibly diverse, including tropical, temperate, desert, arid and subarctic areas. At three million square kilometres (1.2 million square miles), the continent's largest geographical feature is the Sahara Desert. Areas north of the Sahara have had close interactions with Eurasia since the fourth millennium BCE, while the sub-Saharan region was comparatively isolated until the medieval period; Bantu peoples from West Central Africa settled across the southern part of the continent between 2000 BCE to 1000 CE. The transatlantic slave trade was a major destabilizing force in Africa, with more than 10 million Africans transported across the Atlantic between 1500 and 1850. In the late 19th century, European powers carved up the continent between them with little regard for its indigenous peoples. Most countries won independence during the 1950s and 1960s, but decolonization was often traumatic, with warfare, corruption and instability keeping many nations poor and underdeveloped.

Cape Verde

An archipelago of ten, largely arid, principal islands in the mid-Atlantic Ocean, Cape Verde was uninhabited by humans until its discovery by Portugal in 1456. The Portuguese colonized the islands, bringing slaves from Africa; as a result most Cape Verdeans have a mixture of African and European ancestry. Owing to their strategic position, the islands became an important trading post and supply point for shipping, and were highly active in the slave trade. After five centuries of Portuguese rule, the Cape Verdeans demanded independence, which was achieved in 1975.

Under single-party rule until 1990, the country held its first contested elections the year after. It has since become one of the most stable and free democracies in Africa, as well as one of the least corrupt nations in the continent. Despite a lack of natural resources, Cape Verde has achieved high levels of development and standards of living. Even so, many people have emigrated in search of opportunity and, today, the number of Cape Verdeans living abroad is greater than those remaining at home.

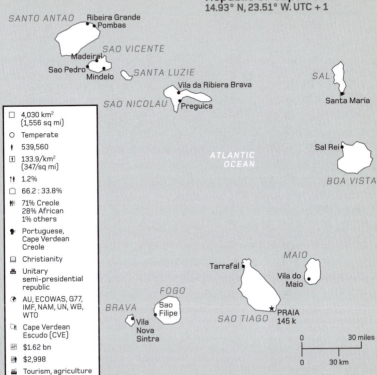

Republic of Cape Verde
14.93° N, 23.51° W. UTC + 1

SANTO ANTAO

Ribeira Grande
Pombas

SAO VICENTE

Madeiral

Sao Pedro
Mindelo

SANTA LUZIE

SAL

Santa Maria

Vila da Ribiera Brava

SAO NICOLAU
Preguica

ATLANTIC
OCEAN

Sal Rei

BOA VISTA

MAIO

Tarrafal

Vila do
Maio

FOGO

BRAVA

Sao
Filipe

Vila
Nova
Sintra

SAO TIAGO

★ PRAIA
145 k

- ☐ 4,030 km² (1,556 sq mi)
- O Temperate
- 539,560
- 133.9/km² (347/sq mi)
- 1.2%
- 66.2 : 33.8%
- 71% Creole 28% African 1% others
- Portuguese, Cape Verdean Creole
- Christianity
- Unitary semi-presidential republic
- AU, ECOWAS, G77, IMF, NAM, UN, WB, WTO
- Cape Verdean Escudo (CVE)
- $1.62 bn
- $2,998
- Tourism, agriculture and food processing

0   30 miles

0   30 km

Senegal

A low-lying country in West Africa, Senegal was dominated by the Wolof Empire until the arrival of Europeans in the mid-15th century, and the Wolof ethnic group still tend to be the leading force in Senegalese political and economic life. France became the main colonial power here, using the country as a base for its slave trade. Although the French encouraged Christian missionaries, Islam – first introduced to the region from North Africa in the mid-11th century – remained the dominant religion. Alongside Mali, Senegal achieved independence in 1960 as part of the Mali Federation, but Senegal withdrew after just a few weeks to become a separate state. In 1982, Senegal and The Gambia began an attempt at integration, forming the Senegambia Confederation, but this union was dissolved in 1989. Conflict between the Senegalese government and separatists in the southern region of Casamance from 1982 onwards ended with a ceasefire in 2014. Senegal has become a stable democracy, with power changing hands in peaceful transitions after the presidential elections of 2000 and 2012.

Republic of Senegal
14.76° N, 17.37° W
UTC

ATLANTIC OCEAN

Senegal

Saint-Louis

Louga

MAURITANIA

DAKAR
3.5 m

Mbour

Diourbel

Kaolack

BANJUL

GAMBIA

Tambacounda

MALI

Gambia

Kolda

GUINEA-BISSAU

GUINEA

☐	196,710 km² (75,950 sq mi)
O	Tropical
⚲	15,411,614
⊞	80.0/km² (207/sq mi)
⇅	2.9%
⌂	44.0 : 56.0%
⚥	39% Wolof 27% Pular 15% Serer 19% others
⚑	Wolof, French
⌺	Islam
♞	Semi-presidential republic
◉	AU, ECOWAS, G-15, G77, IMF, NAM, UN, WB, WTO
⌨	West African CFA Franc (XOF)
▦	$14.76 bn
▨	$958
⛏	Mining (particularly phosphates and limestone), agriculture, tourism

0 ————— 120 miles

0 ————— 120 km

Mauritania

With the majority of its territory lying within the Sahara desert, Mauritania is one of Africa's least densely populated countries. Islam was first introduced in the eighth century, and became the dominant religion when the Muslim Almoravid dynasty conquered the country in the mid-11th century. France gained control of coastal areas in 1817 and gradually extended its influence into the interior, fighting the various emirates that ruled the area and establishing colonial rule that lasted until 1960. Mauritania ruled the southern third of Western Sahara from 1973, before it was forced to withdraw in 1976 by local guerrillas, relinquishing its claim to Morocco. Between 1989 and 1991, the country fought a territorial war with its southern neighbour Senegal that displaced thousands. After decades of authoritarian rule, the country's first multiparty elections were held in 2007. Despite a move towards democracy, however, ethnic divisions remain, with political power and wealth concentrated among the Arab–Berber Sahrawi at the expense of the Haratin (Arabic-speaking descendants of slaves) and sub-Saharan African peoples.

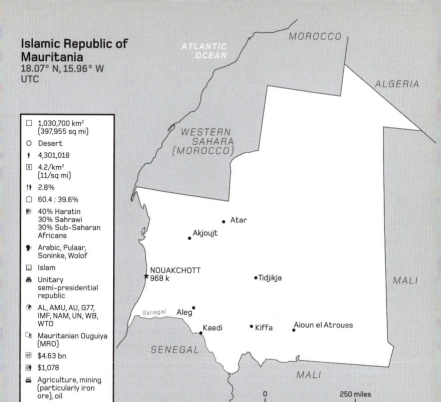

Islamic Republic of Mauritania
18.07° N, 15.96° W
UTC

ATLANTIC
OCEAN

MOROCCO

ALGERIA

WESTERN
SAHARA
(MOROCCO)

MALI

SENEGAL

MALI

- □ 1,030,700 km² (397,955 sq mi)
- ○ Desert
- ✝ 4,301,018
- ▣ 4.2/km² (11/sq mi)
- ↑↑ 2.8%
- ⬭ 60.4 : 39.6%
- ♟ 40% Haratin 30% Sahrawi 30% Sub-Saharan Africans
- ♟ Arabic, Pulaar, Soninke, Wolof
- ▭ Islam
- ⚖ Unitary semi-presidential republic
- ◉ AL, AMU, AU, G77, IMF, NAM, UN, WB, WTO
- ▯ Mauritanian Ouguiya (MRO)
- ▦ $4.63 bn
- ▨ $1,078
- ⬛ Agriculture, mining (particularly iron ore), oil

- Atar
- Akjoujt
NOUAKCHOTT
★ 968 k
- Tidjikja
Senegal
Aleg
- Kaedi
- Kiffa
- Aioun el Atrouss

0   250 miles

0   250 km

The Gambia

The smallest mainland African state, The Gambia is dominated by its eponymous river, which forms a central spine across the country. Before the arrival of Europeans in the mid-15th century, the area was ruled by medieval African monarchies, including the Takrur and the Mali Empire. During the 17th and 18th centuries, France and Britain vied for control of the Gambia River – an important centre of the slave trade.

British rule was officially established in 1821, at first in a small coastal area around the capital, Bathurst (Banjul from 1973), before spreading further inland. The Gambia won independence in 1965 and, despite occasional military coups, the postcolonial regime has been largely stable. In 1982, a treaty of confederation was signed with Senegal, which completely surrounds The Gambia by land; the union, which had little public support, ended in 1989. With few natural resources, the economy remains largely agricultural and heavily reliant on peanut farming, as well as tourism and remittances from Gambians living abroad.

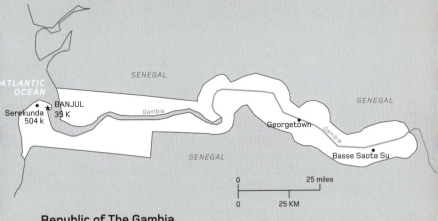

Republic of The Gambia
13.45° N, 16.58° W
UTC

☐ 11,300 km² (4,363 sq mi)	👫 34% Mandinka/Jahanka 22% Fulani/Tukulur/Lorobo 12% Wolof 11% Jola/Karoninka 21% others	🌐 AU, ECOWAS, G77, IMF, NAM, UN, WB, WTO
○ Tropical		💱 Gambian Dalasi (GMD)
👤 2,038,501		📊 $964.60 m
⊞ 201.4/km² (522/sq mi)	🗣 English	📈 $473
↑↑ 3.0%	🕌 Islam	🚜 Agriculture, tourism
☐ 60.2 : 39.8%	⚖ Unitary presidential republic	

Guinea-Bissau

The mostly flat country of Guinea–Bissau has an interior made up of plains and a coast lined with mangrove swamps. From the 13th century, it was part of the Mali Empire that covered much of West Africa. Europeans started to arrive in the 15th century, drawn by the slave trade. The Portuguese established colonial rule, exporting slaves until the 1870s, but were confined to coastal areas, with local rulers dominating the interior. A violent 'pacification' campaign in 1913–15 brought the territory more closely under Portuguese authority. In 1956, nationalists began to campaign for independence from Portugal, first politically but after 1963 through guerrilla warfare. After more than a decade of fighting, independence came in 1974; 'Bissau' was added to the country's name to distinguish it from Guinea, its larger neighbour to the south. Although free and fair elections were held in 2014, instability has been endemic in Bissau–Guinean politics, with periods of military rule, coups and periodic internal violence. The country suffers from high levels of international debt and its economy remains underdeveloped with farming, particularly of cashews, continuing to be vital.

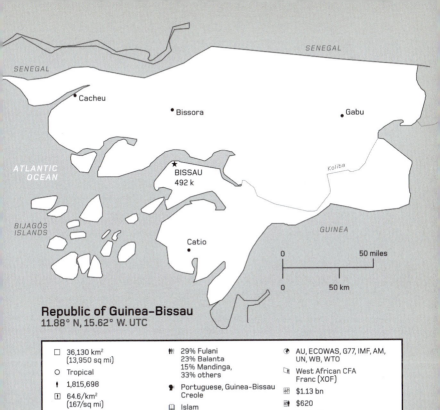

SENEGAL

SENEGAL

• Cacheu

• Bissora

• Gabu

ATLANTIC
OCEAN

Koliba

★ BISSAU
492 k

BIJAGÓS
ISLANDS

GUINEA

• Catio

| 0 | | 50 miles |
| 0 | | 50 km |

Republic of Guinea-Bissau
11.88° N, 15.62° W. UTC

☐	36,130 km² (13,950 sq mi)	♔	29% Fulani 23% Balanta 15% Mandinga, 33% others	☺	AU, ECOWAS, G77, IMF, AM, UN, WB, WTO
O	Tropical				
✝	1,815,698	♟	Portuguese, Guinea-Bissau Creole	☐	West African CFA Franc (XOF)
⊞	64.6/km² (167/sq mi)	☐	Islam	⊞	$1.13 bn
↑↑	2.5%	♣	Unitary semi-presidential republic	⊞	$620
☐	50.1 : 49.9%			☰	Agriculture

Guinea

The Guinean landscape largely consists of coastal plain that gives way to interior highlands. Before colonization, the country was part of a succession of West African empires and kingdoms. European slave traders were active in coastal areas from the 16th century, but it was not until the late 19th century that France began to venture into the interior, defeating local powers to establish the colony of French Guinea. In a 1958 referendum, the public voted to cut ties with France and the country declared itself an independent republic. Nationalist leader Ahmed Sékou Touré became president, declaring his party the only legal one and leading a regime responsible for the death or exile of thousands of people. Since Touré's death in 1984, there have been several military coups and violent clashes between ethnic and political factions, but the country now appears to be moving towards more democratic, civilian rule. Long-term instability has prevented Guinea benefitting from its rich natural resources; it has fertile soil, largely unexploited mineral deposits (bauxite, iron ore and gold) and the potential to generate hydropower from the Gambia, Niger and Senegal rivers.

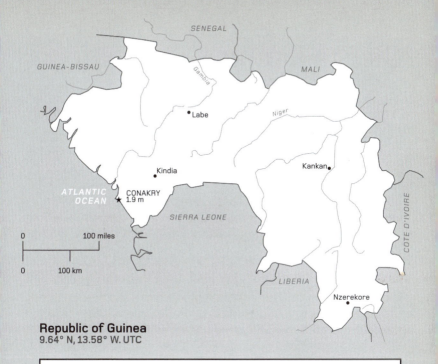

Republic of Guinea
9.64° N, 13.58° W. UTC

☐ 245,860 km² (94,927 sq mi)	☐ 37.7 : 62.3%	✪ AU, ECOWAS, G77, IMF, NAM, UN, WB, WTO
○ Tropical	♛ 34% Fulani 31% Malinke 19% Susu 16% other	▧ Guinean Franc (GNF)
⬩ 12,395,924		▨ \$6.29 bn
⊞ 50.4/km² (131/sq mi)	♟ French	▧ \$508
♛♛ 2.5%	▣ Islam	⬛ Mining (particularly bauxite), oil
	♨ Presidential republic	

Sierra Leone

In 1462, a Portuguese explorer named some hills lying on a peninsula off the West African coast Serra Lyoa, or 'Lion Mountains'. Transliterated into English this became Sierra Leone. By the late 15th century, Europeans had established trading posts there, primarily to purchase slaves. These outposts became the site of Freetown, founded in 1792 as a colony for people of African descent, who had remained loyal to Britain during the American War of Independence. On the abolition of the slave trade in the British Empire in 1807, thousands of former African slaves were resettled here. In 1896, the area around the colony, which extended from the coast inland towards the plains, plateau, mountains and swamps of the interior, became a British protectorate. Together with Freetown, the protectorate won independence in 1961 as modern Sierra Leone. Initially, the country was peaceful and democratic, but a 1967 military coup began a long period of corrupt and repressive rule that culminated in the 1991–2002 civil war. Since peace was made, the country has transitioned back to democracy and begun to rebuild its economy, primarily through diamond and iron-ore mining.

GUINEA

GUINEA

Makeni ●

Rokel

Magburaka ●

Sefadu ●

★ FREETOWN
1 m

● Bo

LIBERIA

ATLANTIC
OCEAN

Moa

☐	72,300 km² (27,915 sq mi)
○	Tropical
✝	7,396,190
⊞	102.5/km² (266/sq mi)
↑↑	2.2%
☐	40.3 : 59.7%
👪	35% Temne 31% Mende 34% others
👤	Krio, English
📖	Islam
⚖	Unitary presidential constitutional republic
🌐	AU, CON, ECOWAS, G77, IMF, NAM, UN, WB, WTO
💱	Sierra Leonean Leone (SLL)
📊	$3.67 bn
📈	$496
⛏	Mining (particularly minerals), agriculture

Republic of Sierra Leone
8.47° N, 13.23° W
UTC

0   50 miles

0   50 km

Morocco

Separated from Europe by less than 14 kilometres (9 miles), Morocco has a landscape mainly comprising desert and mountains, with a coastline on both the Mediterranean Sea and the Atlantic Ocean. At times part of the Carthaginian, Roman and Byzantine empires, since the eighth century the country has been ruled by a series of Muslim dynasties culminating in the current royal house, the Alaouites, who came to power in 1666. A French protectorate was declared in 1912, with Spain ruling an enclave to the south of the country and the northern coast (with the exception of the 'international zone' of Tangier). Foreign rule ended in 1956, after which the country became an independent constitutional monarchy. The king retained significant power and, until the 1980s, political dissidents were subject to violent state suppression. Protests demanding reform in 2011–12 led to a new constitution that slightly limited royal authority and increased the power of parliament. When Spain withdrew from Western Sahara in 1976, Morocco annexed the territory, although its sovereignty is not internationally recognized.

ATLANTIC OCEAN

MEDITERRANEAN SEA

SPAIN

ATLANTIC OCEAN

CANARY ISLANDS (SPAIN)

ALGERIA

MAURITANIA

WESTERN SAHARA

0 — 200 miles
0 — 200 km

RABAT 2.0 m ★
Kenitra
Casablanca 3.5 m
El Jadida
Safi
Essaouira
Marrakech
Agadir
Bou Izakarn
Goulimine
Tan-Tan
Laayoune
Dakhla

Meknes
Khouribga
Kasba Tadla
Beni Mellal
Oued Zem
Boumaine-Dades
Tazenakht

Fes
Oujda
Berguent
Figuig

Sebou
Moulouya
Draa

M O U N T A I N S

A T L A S

Kingdom of Morocco
33.97° N, 6.85° W. UTC

Western Sahara was annexed by Morocco following Spanish withdrawal in 1976. The Sahrawi Arab Democratic Republic claims to represent the territory and administers parts of it as an independent state.

☐ 446,550 km² (172,414 sq mi)	♟ Unitary parliamentary system under constitutional monarchy
○ Mediterranean, desert in the interior	⊕ AL, AMU, AU, G77, IMF, NAM, UN, WB, WTO
♦ 35,276,786	🗐 Moroccan Dirham (MAD)
⊞ 79.0/km² (205/sq mi)	🖾 $101.45 bn
♦♦ 1.4%	🖳 $2,832
◻ 60.7 : 39.3%	♨ Mining (particularly phosphates), agriculture, textiles
♛♛ 99% Arab-Berber 1% others	
♠ Arabic, Berber	
▦ Islam	

Mali

A mostly flat, dry and landlocked country in West Africa, Mali has been ruled by three of the continent's greatest historical powers: the Ghana, Mali and Songhai empires. In the 13th–16th centuries, the area played a crucial role in trans-Saharan commerce and was one of the largest gold producers in the world, as well as an important centre of Islam. This period of prosperity, largely built on the gold trade, came to an end with the establishment of sea routes between Europe and Africa, coupled with military defeats by Morocco. French colonial rule began in the late 19th century, and lasted until Mali won independence in 1960 – initially as part of a short-lived federation with Senegal. Radical socialist reforms led to a 1968 military coup that installed Moussa Traoré as president, a dictator who was overthrown in 1991. Two decades of stability followed, with four consecutive multiparty elections, but this came to an end in 2012 when ethnic Tuareg separatists in the north rebelled and army officers overthrew the president. Since this upheaval, the rebellion has been largely defeated and democratic rule re-established.

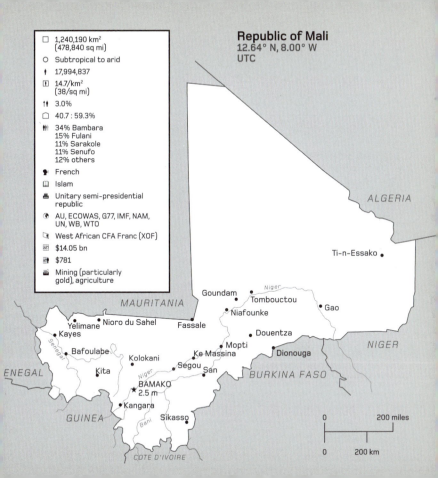

Republic of Mali
12.64° N, 8.00° W
UTC

□ 1,240,190 km²
(478,840 sq mi)

○ Subtropical to arid

† 17,994,837

⊞ 14.7/km²
(38/sq mi)

†† 3.0%

◯ 40.7 : 59.3%

††† 34% Bambara
15% Fulani
11% Sarakole
11% Senufo
12% others

🕭 French

📖 Islam

♣ Unitary semi-presidential
republic

◉ AU, ECOWAS, G77, IMF, NAM,
UN, WB, WTO

▧ West African CFA Franc (XOF)

▨ $14.05 bn

▧ $781

▨ Mining (particularly
gold), agriculture

ALGERIA

Ti-n-Essako •

MAURITANIA

Niger

Goundam •   Tombouctou • Gao
  • Niafounke

Fassale   Douentza
Yelimane • Nioro du Sahel   •

Kayes •   Mopti
                 •
Bafoulabe •   Kolokani   Ke Massina   Dionouga
                         •
  Kita •   Segou •   NIGER
            Niger   San •
                        BURKINA FASO
   BAMAKO ★
   2.5 m

Kangara •
                Bani
         Sikasso •

SENEGAL

GUINEA

COTE D'IVOIRE

0   200 miles

0   200 km

Liberia

In 1822, a colony for freed African-American slaves was founded in an area known as the Grain Coast, after the melegueta pepper grains that grew and were traded there. The territory, now named Liberia for the liberty its inhabitants hoped to enjoy, declared independence from its American founders in 1847, with a democratic constitution based on that of the United States; it is the oldest republic in Africa. The settlers and their descendants, known as Americo-Liberians, inhabited coastal areas, while indigenous chiefdoms dominated the interior. Although a minority, the Americo-Liberians became the country's dominant force. Their True Whig Party held power from 1878–1980, when local ethnic groups overthrew it in a military coup. Subsequently, the country suffered bloody civil wars from 1989–97 and 1999–2003. Free and fair elections held in 2005 established democratic rule but, despite this political progress, the country faces significant challenges, particularly regarding issues of corruption, inadequate infrastructure and illiteracy. The Liberian economy has begun to rebuild, primarily through mining of resources such as iron ore, diamonds and gold.

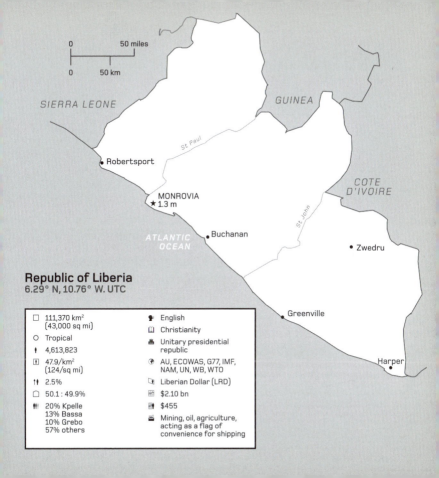

SIERRA LEONE

GUINEA

COTE D'IVOIRE

St Paul

St John

• Robertsport

★ MONROVIA
1.3 m

ATLANTIC
OCEAN

• Buchanan

• Zwedru

• Greenville

Harper •

Republic of Liberia
6.29° N, 10.76° W. UTC

☐ 111,370 km² (43,000 sq mi)	♟ English
◯ Tropical	📖 Christianity
⚥ 4,613,823	♟ Unitary presidential republic
⊞ 47.9/km² (124/sq mi)	🌐 AU, ECOWAS, G77, IMF, NAM, UN, WB, WTO
⚦ 2.5%	💵 Liberian Dollar (LRD)
☐ 50.1 : 49.9%	📊 $2.10 bn
♟ 20% Kpelle 13% Bassa 10% Grebo 57% others	📈 $455
	🚢 Mining, oil, agriculture, acting as a flag of convenience for shipping

Algeria

Algeria might be Africa's largest country, but its population is concentrated in a relatively small area along its Mediterranean coast, owing to the harsh desert climate of the interior. Ruled at times by the Carthaginians, Romans, Arabs and Ottomans, local tribes always maintained a high degree of importance and autonomy. France invaded Algeria in 1830, imposing colonial rule and encouraging the migration of hundreds of thousands of Europeans. The indigenous Muslim population was marginalized, leading to a bloody war of liberation from 1954–62. The conflict ended with Algeria gaining independence and one million Europeans fleeing the country. After more than three decades of single-party socialist rule, open elections in 1991 saw Islamists win a majority in the legislature. Concerned that they might gain power, the army cancelled voting, sparking a decade-long civil war between the government and Islamist rebels. Government forces triumphed and there has been a slow restoration of stability and democracy, with liberal reforms introduced in the aftermath of the 2011 regional protests known as the 'Arab Spring'.

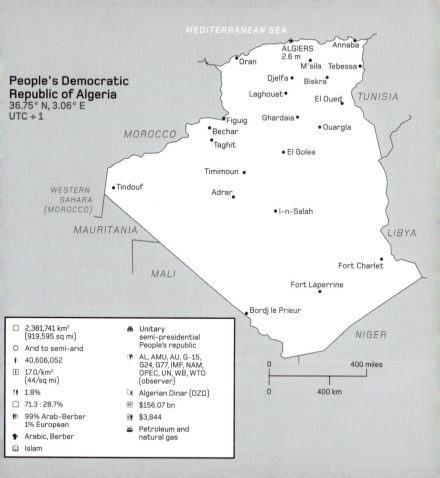

MEDITERRANEAN SEA

People's Democratic Republic of Algeria
36.75° N, 3.06° E
UTC +1

MOROCCO

WESTERN
SAHARA
(MOROCCO)

MAURITANIA

MALI

ALGIERS
2.6 m

Annaba

Oran

M'sila
Tebessa

Djelfa
Biskra

Laghouat

El Oued

TUNISIA

Ghardaia

Figuig

Ouargla

Bechar

Taghit

El Golea

Timimoun

Tindouf

Adrar

I-n-Salah

LIBYA

Fort Charlet

Fort Laperrine

Bordj le Prieur

NIGER

▢ 2,381,741 km² (919,595 sq mi)	♖ Unitary semi-presidential People's republic
ⵔ Arid to semi-arid	
⚲ 40,606,052	⊕ AL, AMU, AU, G-15, G24, G77, IMF, NAM, OPEC, UN, WB, WTO (observer)
⊞ 17.0/km² (44/sq mi)	
♁ 1.8%	⛃ Algerian Dinar (DZD)
▢ 71.3 : 28.7%	⛁ $156.07 bn
ⵔ 99% Arab-Berber 1% European	⛁ $3,844
♟ Arabic, Berber	⛃ Petroleum and natural gas
⛩ Islam	

0 ——————— 400 miles

0 ——————— 400 km

Côte d'Ivoire

Named by Europeans for its major export at the time, the West African nation of Côte d'Ivoire (Ivory Coast) extends from its Atlantic seaboard to an interior of forests and savannah. During the late 19th century, France established trading posts along the coast, and after declaring the territory a colony in 1893, extended control into the interior, imposing a plantation system based on forced labour. The colonial economy relied on the export of cash crops, such as coffee, cocoa and palm oil. Agriculture continued to be essential after independence in 1960, and helped the country to achieve a long period of sustained economic growth. The country transitioned from single-party rule to full democracy in 1990, but a 1999 military coup led to a civil war (2002–7) between Muslim rebels in the north and mostly Christian government forces in the south. The conflict was renewed in 2010, when sitting president Laurent Gbagbo refused to hand over power after losing elections. After four months of fighting, he was ousted by supporters of the victorious candidate. Subsequently, the new regime attempted to restore harmony to a divided nation.

Republic of
Côte d'Ivoire
6.83° N, 5.29° W
UTC

☐	322,400 km² (124,503 sq mi)
○	Tropical along coast, semi-arid in far north
☩	23,695,919
⊞	74.5/km² (193/sq mi)
⇈	2.5%
◠	54.9 : 45.1%
⋔	29% Akan 16% Voltaique/Gur 15% Northern Mande 40% others
♟	French
⌨	Islam, Christianity
♣	Unitary presidential republic
☯	AU, ECOWAS, G24, G77, IMF, NAM, UN, WB, WTO
⌦	West African CFA Franc (XOF)
▩	$36.16 bn
▦	$1,526
▦	Agriculture

MALI

BURKINA FASO

GUINEA

Korhogo

Lake Kossou

Bouake

Daloa

YAMOUSSOUKRO
★ 259 k

Dimbokro

GHANA

Gagnoa

Sassandra

Bandama

LIBERIA

Abidjan
4.9 m

GULF OF GUINEA

0   200 miles

0   200 km

Burkina Faso

The landlocked western African nation of Burkina Faso lies on a large plateau that features arid savannah to the north and forested areas to the south. By the 15th century, the Mossi people had established several kingdoms there. When France claimed the territory in 1898, its colonial forces faced concerted local resistance, and indigenous rulers continued to be influential. The country won independence as Upper Volta in 1960, but the ruling regime was overthrown in a military coup six years later, beginning a period of sustained political instability. In 1983, Thomas Sankara, a young army officer, won power and instituted a series of ambitious reforms such as promoting women's rights, encouraging vaccination and fighting corruption. As an expression of this new regime, the country was renamed Burkina Faso (Land of Incorruptible People). After less than four years, Sankara was overthrown by a former ally, Blaise Compaoré, who reversed most of these policies. Compaoré ruled until 2014, when popular protest forced him to resign. Although multiparty elections were held in 2015, the country remains poor and largely undeveloped.

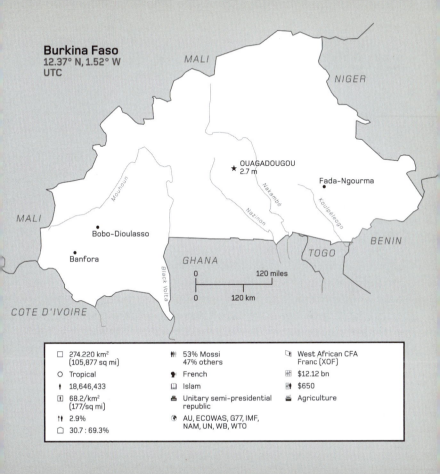

Burkina Faso
12.37° N, 1.52° W
UTC

MALI

NIGER

MALI

★ OUAGADOUGOU
2.7 m

Fada-Ngourma

Mouhoun

Nazinon

Natamba

Koulpeleogo

Bobo-Dioulasso

Banfora

GHANA

Black Volta

TOGO

BENIN

COTE D'IVOIRE

| 0 | 120 miles |
| 0 | 120 km |

☐ 274.220 km² (105,877 sq mi)	🚶 53% Mossi 47% others	💱 West African CFA Franc (XOF)
○ Tropical	🗣 French	💹 $12.12 bn
♀ 18,646,433	🕌 Islam	💰 $650
▦ 68.2/km² (177/sq mi)	⚖ Unitary semi-presidential republic	🚜 Agriculture
♀♂ 2.9%	🌍 AU, ECOWAS, G77, IMF, NAM, UN, WB, WTO	
☐ 30.7 : 69.3%		

Ghana

Lying on the Gulf of Guinea in West Africa, Ghana is home to Lake Volta, the world's largest artificial lake. From the 11th century, Ghanaian territory was ruled by local kingdoms, the most powerful of which was the Ashanti, with a wealth and prominence based on the gold trade. Drawn primarily by that gold, Europeans began to arrive in the late 15th century, establishing numerous trading posts along the coast. By the 19th century, Britain had become the dominant imperial force, fighting a series of wars with the Ashanti to maintain control, and creating the colony of the Gold Coast. In 1957, the country was the first former colony in sub-Saharan Africa to gain independence, taking the name Ghana. Its first president, Kwame Nkrumah, was a visionary socialist leader and proponent of African unity, but was overthrown in a 1966 military coup. The country entered a long period of political instability, and multiparty elections were not held until 1992. Democratic rule has now been successfully established and economic development has been robust, thanks to exports of gold and cocoa and a burgeoning oil and natural gas sector.

Republic of Ghana
5.60° N, 0.19° W
UTC

- ☐ 238,540 km² (92,101 sq mi)
- ○ Tropical
- ⚲ 28,206,728
- ⊞ 124.0/km² (321/sq mi)
- ⚧ 2.2%
- ◗ 54.7 : 45.3%
- ⚦ 47% Akan
  17% Mole-Dagbon
  14% Ewe
  22% other
- ⚲ English
- ⚏ Christianity
- ⛏ Unitary presidential constitutional republic
- ◉ AU, CON, ECOWAS, G24, G77, IMF, NAM, UN, WB, WTO
- ⚲ Ghana Cedi (GHS)
- ▦ $42.69 bn
- ▦ $1,513
- ⛏ Mining (particularly gold), oil and natural gas, agriculture

BURKINA FASO

Black Volta

• Wa

White Volta

• Tamale

COTE D'IVOIRE

TOGO

Lake Volta

• Kumasi
2.6 m

Obuasi

ACCRA ★
2.3 m

KETA

Takoradi •

GULF OF GUINEA

0 ___ 60 miles

0 ___ 60 km

Togo

A narrow West African country, Togo measures just 140 km (87 miles) east to west at its widest point. Settled by various ethnic groups, including the Ewe from whose language its name derives, it was an important centre of the transatlantic slave trade. In 1884, Germany established a protectorate here, which British and French forces captured in the First World War; France governed the east and Britain the west. In 1957, following a referendum, British Togoland became part of Ghana. French Togoland won independence as the Togolese Republic in 1960, and in 1967 Gnassingbé Eyadéma became president following a military coup. Opposition parties were banned until 1993, although Eyadéma's dictatorial rule continued until his death in 2005. Installed by the army as president, his son Faure Gnassingbé consolidated power by winning elections held that year. Opposition parties denounced their legitimacy, however, and factional clashes ensued. Free and fair elections were held in 2007, and Togo is moving towards political stability. Recent economic growth has also been steady, thanks to the country's phosphate deposits and the export of cash crops, such as cotton and coffee.

Togolese Republic
6.17° N, 1.23° E
UTC

☐	56,790 km² (21,927 sq mi)
○	Tropical in south, semi-arid in north
⋔	7,606,374
▦	139.8/km² (362/sq mi)
⋔⋔	2.5%
⌂	40.5 : 59.5%
⋔⋔⋔	22% Ewe 13% Kabre 10% Wachi 55% other
♟	French, Ewe, Kabiyé
▢	Traditional indigenous beliefs
▣	Presidential republic
◉	AU, ECOWAS, G77, IMF, NAM, UN, WB, WTO
▢	West African CFA Franc (XOF)
▦	$4.39 bn
▦	$578
▤	Agriculture, mining (particularly phosphates)

BURKINA FASO

• Dapaong

BENIN

• Mango

• Niamtougou

• Kara

GHANA

• Tchamba

Sokode •

• Sotouboua

• Atakpame

BENIN

LOME
956 k
★

BIGHT OF BENIN

Lake Volta

Oti

Mono

0 — 50 miles
0 — 50 km

Niger

Named after the longest river in West Africa, most of Niger is covered by the Sahara desert, with the majority of its population living in the comparatively more hospitable south and west of the country. The Songhai, Mali and Kanem-Bornu empires ruled the area before France claimed it as part of its sphere of influence in 1885. To impose colonial rule, the French launched several military expeditions that were characterized by their violent and cruel treatment of indigenous civilians. Independence was acquired in 1960, and for the next 14 years the country was under single-party civilian rule. Political repression continued after a 1974 military coup, and free elections were not held until 1993. However, the elected president was overthrown in 1996, triggering a period of political instability that lasted until 2011. Allied to its political upheavals, Niger has extremely low levels of development and the lowest literacy rate in the world. Despite having potentially valuable deposits of uranium and other minerals, the economy is largely based on subsistence agriculture, and drought remains a major problem.

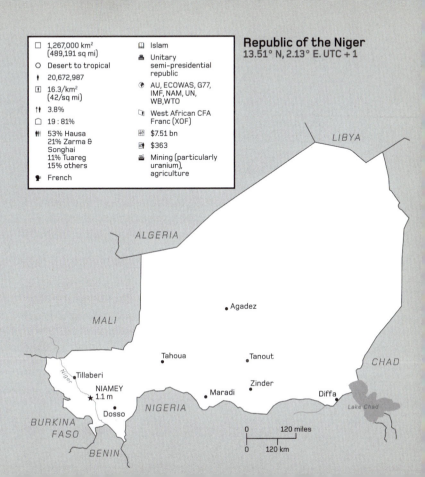

Republic of the Niger
13.51° N, 2.13° E. UTC + 1

□ 1,267,000 km²
(489,191 sq mi)

○ Desert to tropical

† 20,672,987

⊡ 16.3/km²
(42/sq mi)

†† 3.8%

⌂ 19 : 81%

††† 53% Hausa
21% Zarma &
Songhai
11% Tuareg
15% others

🐾 French

📖 Islam

♨ Unitary
semi-presidential
republic

◉ AU, ECOWAS, G77,
IMF, NAM, UN,
WB,WTO

💱 West African CFA
Franc (XOF)

▦ $7.51 bn

▨ $363

⚒ Mining (particularly
uranium),
agriculture

LIBYA

ALGERIA

MALI

• Agadez

Tahoua •   • Tanout

Niger • Tillaberi
★ NIAMEY
   1.1 m   Zinder •
              • Maradi   Diffa •
• Dosso   Lake Chad

BURKINA   NIGERIA
FASO

CHAD

BENIN

0   120 miles
0   120 km

Benin

The West African state of Benin extends over 640 km (400 miles) from the Niger River in the north to the Atlantic coast in the south. Prior to the late 19th century, Beninese territory was dominated by local kingdoms, the most important of which, the coastal state of Dahomey, was highly involved in selling slaves to European traders. During the 1890s, France added the area to its empire as French Dahomey. It was granted independence in 1960, but rivalry between different ethnic groups made the new state highly unstable. In 1972, the government was overthrown and single-party Marxist–Leninist rule was declared; three years later, the state was renamed Benin. Protests in 1989 led the government to liberalize, and elections were held in 1991. Since then, Benin has transitioned from a dictatorship to a multiparty democracy. It has also moved towards a free-market economic system, with cotton being the most important source of export income. However, much of the population is still employed in subsistence agriculture, and corruption and inadequate infrastructure remain long-term challenges for the government to address.

Republic of the Benin
6.50° N, 2.63° E
UTC + 1

☐	114,760 km² (44,300 sq mi)
○	Tropical
⚲	10,872,298
⊞	96.4/km² (250/sq mi)
↑↑	2.8%
◡	44.4 : 55.6%
⚶	38% Fon and related 15% Adja and related 12% Yoruba and related 10% Bariba and related 25 % others
⚑	French
📖	Christianity
♟	Presidential republic
⊕	AU, ECOWAS, G77, IMF, NAM, UN, WB, WTO
💵	West African CFA Franc (XOF)
▦	$8.58 bn
▥	$789
⛏	Agriculture (particularly cotton)

Nigeria

Africa's most populous country, Nigeria lies to the west of the continent and mostly consists of plains with areas of highland and plateaus in the centre. It is a multinational state with hundreds of ethnic groups, broadly split between followers of Christianity in the south and Islam in the north. The area was a major centre of the slave trade, leading Europeans to establish outposts in coastal areas. British rule formally began in 1901, and continued until 1960. Nigeria then adopted a federal system of government where local areas retained some autonomy; initially the country had three regions, but now it has 36 states as well as a territory around Abuja, the capital since 1991. Stability has been hard to achieve; between 1967 and 1970 the southeastern region of Biafra fought an unsuccessful war of secession against the federal government, and until 1999 the country was ruled by largely corrupt military juntas. Nigeria has since transitioned to democratic civilian rule, but has faced Islamist violence led by Boko Haram rebels since 2002. Despite its political difficulties Nigeria has the largest economy in Africa, largely based on its oil industry.

Federal Republic of Nigeria
9.08° N, 7.40° E. UTC + 1

☐ 923,770 km² (356,679 sq mi)	👪 29% Hausa and Fulani 21% Yoruba 18% Ibo 10% Ijaw 22% other	⚙ AU, CON, ECOWAS, G-15, G24, G77, IMF, NAM, OPEC, UN, WB, WTO
O Arid in north, tropical in centre, equatorial in south		
⬍ 185,989,640	👤 English	🗋 Nigerian Naira (NGN)
⊞ 204.2/km² (529/sq mi)	🕮 Christianity, Islam	▦ $405.08 bn
⇈ 2.6%	🖥 Federal presidential republic	▤ $2,178
◲ 48.6 : 51.4%		▥ Oil

Equatorial Guinea

Comprising a mainland territory on the Central African coast and five inhabited islands, Equatorial Guinea was peopled by various ethnic groups, including the Fang, before coming under Spanish rule in 1778. The Spanish established a plantation economy based on cocoa and coffee production, but attempts to increase profitability were hampered by disease and a shortage of labour.

Following independence in 1968, the first elected ruler, Francisco Macías Nguema, named himself president for life. He censored the press and oversaw the execution of thousands of civilians. His destructive and violent regime ended in 1979, with a coup organized by his nephew Teodoro Obiang Nguema Mbasogo, who ordered Nguema's execution. However, Mbasogo's ongoing presidency is also authoritarian and corrupt. In 1995, oil was discovered offshore, leading to the country becoming one of Africa's fastest-growing economies. This has not delivered prosperity or progress, however, as revenues are very unevenly distributed, with most reportedly benefiting the president and his family.

MALABO
187 k

Luba

GULF OF GUINEA

☐	28,050 km² (10,830 sq mi)	📖	Christianity
○	Tropical	♟	Dominant-party presidential republic
⚥	1,221,490	⊕	AU, ECCAS, G77,IMF, NAM, OPEC, UN, WB, WTO (observer)
⊞	43.5/km² (113/sq mi)		
⇈	3.8%	🏦	Central African CFA Franc (XAF)
◐	40.1 : 59.9%	📊	$10.18 bn
⚵	86% Fang 14% others	📈	$8,333
⚲	Spanish, French, Portuguese	⛽	Oil

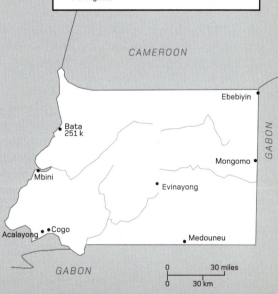

CAMEROON

Ebebiyin

Republic of Equatorial Guinea
3.75° N, 8.74° E
UTC +1

Bata
251 k

GABON

Mongomo

Mbini

Evinayong

Acalayong • Cogo

Medouneu

GABON

0	30 miles
0	30 km

São Tomé and Príncipe

A pair of islands in the Atlantic Ocean, São Tomé and Príncipe are some 145 km (90 miles) apart. They were uninhabited until their discovery by Portuguese explorers around 1470. By 1500, both islands were inhabited by migrants from Portugal and African slaves.

During the 16th century, São Tomé and Príncipe became the world's leading producer of sugar, but soon faced competition from Brazil and the West Indies. The economy shifted towards the slave trade, before concentrating on coffee and cocoa production from the 19th century. Cultivation took place on Portuguese-owned plantations that used indentured labour; conditions were harsh and the colonial regime was repressive and violent. This contributed to a growing nationalist movement, which won independence for São Tomé and Príncipe in 1975. At first the country was under single-party rule, but political reform led to free elections taking place in 1991 and, in 1995, Príncipe was granted autonomy. Despite occasional periods of instability, the process of democratization has been successful although the economy continues to be reliant on cocoa exportation.

Santo Antonio •

PRÍNCIPE

Democratic Republic of São Tomé and Príncipe
0.33° N, 6.73° E
UTC

☐	960 km² (371 sq mi)	⚒	Unitary semi-presidential republic
○	Tropical	◉	AU, ECCAS, G77, IMF, NAM, UN, WB, WTO (observer)
♂	199,910		
⬓	208.2/km² (539/sq mi)	☖	São Tomé and Príncipe Dobra (STD)
♁	2.2%	〽	$351.05 m
◠	65.6 : 34.4%	〽	$1,756
♙	80% Mestiço 10% Fang 10% other	⚍	Agriculture (particularly cocoa)
♟	Portuguese		
▣	Christianity		

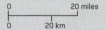

GULF OF GUINEA

0	20 miles
0	20 km

Neves •

★ SÃO TOMÉ
57 k

SÃO TOMÉ

• Santa Cruz

• Porto Alegre

Tunisia

Lying on North Africa's Mediterranean coast, Tunisia was the centre of the Carthaginian empire, a great regional power from the seventh to second centuries BCE. The area then passed under Roman, Vandal, Byzantine, Arab and Ottoman rule. In 1881, France invaded, forcing Tunisia to become a protectorate. Pressure from nationalist groups led to independence in 1956, with Habib Bourguiba taking power as the first president. He ruled for 30 years, and established a one-party system that continued under his successor Zine El Abidine Ben Ali. This secular regime promoted women's rights, but also repressed its political opponents. In 2010–11, during the Arab Spring, a wave of popular protests against the government swept the country; within a month Ben Ali had fled Tunisia. A government of national unity drafted a new constitution, and in 2014 the first free elections since independence were held. Tunisian economic growth has been steady since the 1990s, with the country benefiting from trade with the EU and tourism, although the latter has declined after Islamist terrorist attacks in 2015.

Republic of Tunisia
36.81° N, 10.18° E
UTC + 1

MEDITERRANEAN SEA

Banzart

★ TUNIS
2.0 m

Mejerda

ALGERIA

Sousse

Sfax

GOLFE DE GABÉS

Gabes

Hawmat as Suq

LIBYA

☐	163,610 km² (63,170 sq mi)
○	Temperate in north, desert in south
☥	11,403,248
⊞	73.4/km² (190/sq mi)
↟↟	1.1%
⌂	67.0 : 33.0 %
⚒	98% Arab 2% other
☻	Arabic
📖	Islam
♣	Unitary semi-presidential republic
⊕	AL, AMU, AU, G77, IMF, NAM, UN, WB, WTO
☎	Tunisian Dinar (TND)
▦	$42.06 bn
▦	$3,689
⚒	Oil, mining (particularly phosphates), tourism, textiles

0 ────── 100 miles

0 ────── 100 km

Gabon

arly inhabitants of Gabon were the Babongo, a pygmy people who, by about 1100, had settled the tropical rainforests that comprise over three-quarters of the country. From the 14th century, Bantu tribes from Central Africa migrated into the area. Portuguese explorers arrived in 1472, drawing Gabon into the transatlantic slave trade. France established settlements along the coast during the late 19th century and began exploring the interior. They sold trading concessions to private companies, whose harsh use of forced labour disrupted the indigenous society and economy. In 1910, Gabon became part of French Equatorial Africa.

Colonial rule ended in 1960; since independence the Gabonese Democratic Party has dominated, often with the support of the French government. The country's economy depends on oil, which has been exported since the 1970s. There are also rich deposits of manganese, but despite these natural resources, over one-third of the country lives in poverty owing to government mismanagement and corruption.

Gabonese Republic
0.42° N, 9.47° E
UTC + 1

CAMEROON
Bitam
Oyem
EQUATORIAL GUINEA
REP. OF CONGO
LIBREVILLE
707 k
Ogooué
Lambaréné
Koulamoutou
Omboué
Mouila
Franceville
N'dendé
Setté
Tchibanga
ATLANTIC OCEAN
REP. OF CONGO
0
100 miles
0
100 km

- 267,670 km² (103,348 sq mi)
- Tropical
- 1,979,786
- 7.7/km² (20/sq mi)
- 2.5%
- 87.4 : 12.6%
- 90% Bantu tribes 10% others
- French
- Christianity
- Dominant-party presidential republic
- AU, ECCAS, G24, G77, IMF, NAM, OPEC, UN WB, WTO
- Central African CFA Franc (XAF)
- $14.21 bn
- $7,179
- Oil, manganese mining

Cameroon

One of Africa's most culturally and geographically diverse countries, Cameroon has a landscape that varies from wet coastal forest to desert and is home to over 200 languages. The area played a significant role in trading slaves with both Arabs to the north and Europeans in coastal areas. Germany claimed the territory in 1884, expanding its authority inland and forcing indigenous people to work on its plantations. Following Germany's defeat in the First World War, France and Britain split the colony between them. The French part won independence in 1960 and united with the British part a year later. Cameroon was a single-party state until 1990, and the ruling political faction has always remained in power. Oil reserves were discovered during the 1970s, and the profits have since contributed to steady economic and developmental growth. Despite this, poverty and endemic corruption continue to plague the country. Furthermore, since the 1990s there have been secessionist protests in English-speaking regions to the south, in response to increasing government centralization.

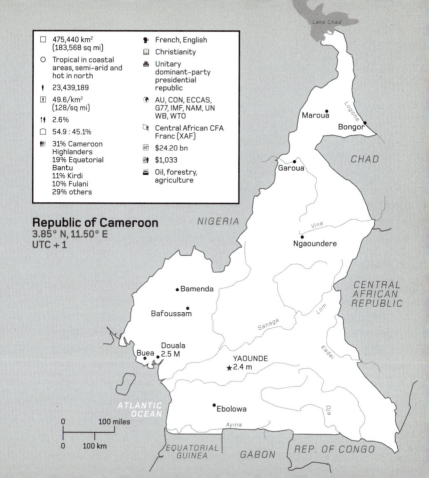

□ 475,440 km²
(183,568 sq mi)

○ Tropical in coastal
areas, semi-arid and
hot in north

♀ 23,439,189

⊡ 49.6/km²
(128/sq mi)

↟ 2.6%

⌂ 54.9 : 45.1%

⚶ 31% Cameroon
Highlanders
19% Equatorial
Bantu
11% Kirdi
10% Fulani
29% others

☙ French, English

⌨ Christianity

♨ Unitary
dominant-party
presidential
republic

⌾ AU, CON, ECCAS,
G77, IMF, NAM, UN
WB, WTO

⌦ Central African CFA
Franc (XAF)

▦ $24.20 bn

▤ $1,033

▭ Oil, forestry,
agriculture

Republic of Cameroon
3.85° N, 11.50° E
UTC + 1

Lake Chad

Maroua
Bongor
Logone

CHAD

Garoua

NIGERIA
Vina

Ngaoundere

CENTRAL
AFRICAN
REPUBLIC

Bamenda

Bafoussam

Lom

Sanaga

Buea
Douala
2.5 M

YAOUNDE
★ 2.4 m

Kadei

Ebolowa

Dja

ATLANTIC
OCEAN

Ayina

0   100 miles
0   100 km

EQUATORIAL
GUINEA   GABON   REP. OF CONGO

Libya

With territory lying predominantly within the Sahara Desert, Libya is most populated along its Mediterranean coast. Its indigenous people are the Berbers, although since the seventh century BCE the area has experienced the rule of other powers, including the Phoenicians, Greeks, Persians, Carthaginians, Romans, Byzantines, Arabs and Ottomans. Italy annexed Libya after winning the Italo-Turkish War (1911–12), but was ousted by the Allies during the Second World War, and the country fell under Franco-British administration. In 1951, Libya declared independence as a constitutional monarchy that was overthrown in a 1969 military coup led by Muammar al-Gaddafi. The new regime combined Islam with socialism; the nationalization of oil brought revenues that increased prosperity and financed major infrastructure projects. Gaddafi also used state money to fund international terrorism. In 2011, Arab Spring protests led to the Libyan Civil War; rebels defeated government forces and Gaddafi was killed. The country descended into another civil war in 2014, and fighting between rival parliaments, local militias and radical Islamists is ongoing.

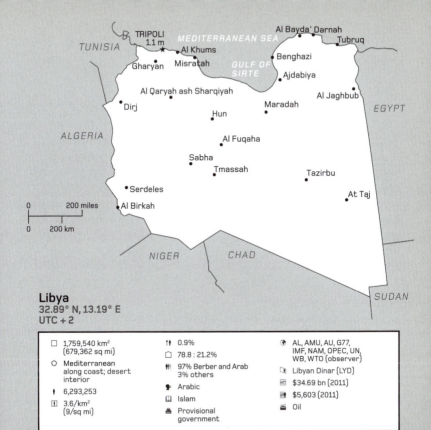

TUNISIA

TRIPOLI
1.1 m

MEDITERRANEAN SEA

Al Bayda' Darnah
Tubruq

Al Khums

GULF OF
SIRTE

Benghazi

Gharyan

Misratah

Ajdabiya

Al Qaryah ash Sharqiyah

Al Jaghbub

EGYPT

Dirj

Maradah

Hun

ALGERIA

Al Fuqaha

Sabha

Tmassah

Tazirbu

Serdeles

At Taj

Al Birkah

0   200 miles

0   200 km

NIGER   CHAD

SUDAN

Libya
32.89° N, 13.19° E
UTC + 2

☐ 1,759,540 km² (679,362 sq mi)	↟ 0.9%	☀ AL, AMU, AU, G77, IMF, NAM, OPEC, UN, WB, WTO (observer)
○ Mediterranean along coast; desert interior	⌂ 78.8 : 21.2%	
	⚥ 97% Berber and Arab 3% others	⛁ Libyan Dinar (LYD)
✝ 6,293,253		▦ $34.69 bn (2011)
⊞ 3.6/km² (9/sq mi)	♟ Arabic	▦ $5,603 (2011)
	⚏ Islam	⛏ Oil
	⚎ Provisional government	

Republic
of the Congo

Named after the Congo River, which runs along its southern border, this Central African country's population descends primarily from the Bantu tribes that had migrated to the area by 1500 BCE. Of the Bantu kingdoms that developed, the most powerful were the Kongo, Loango and Tio. During the 17th and 18th centuries, coastal areas were highly involved in the transatlantic slave trade, and the Congo River became an important route into the interior. In 1880, France claimed the area north of the river after signing a treaty with a local ruler; in 1891, they formally declared it the colony of French Congo.

The country won independence in 1960. Three years later an uprising instituted socialist rule, which ended after multi-party elections were held in 1992. Rivalry between ethnic and political factions led to civil wars in 1993–94 and 1997–99, although stability has largely been restored. The majority of the population continues to live in poverty, despite the country being one of Africa's major producers of oil.

CENTRAL AFRICAN REP.

CAMEROON

Ouesso

Owando

GABON

0   120 miles
0   120 km

DEM REP OF
THE CONGO

Djambala

BRAZZAVILLE
1.9 m

Loubomo

Kinkala

Pointe-Noire

ATLANTIC
OCEAN

**Republic
of the Congo**
0.23° S, 15.83° E
UTC + 1

☐ 342,200 km²
(132,047 sq mi)

O Tropical

👤 5,125,821

▥ 15/km²
(39/sq mi)

👥 2.6%

⬭ 65.8 : 34.2%

👪 48% Kongo
20% Sangha
17% Teke
12% M'Bochi
3% others

🗣 French

📖 Christianity

🏛 Unitary
semi-presidential
republic

🌐 AU, ECCAS, G77, IMF,
NAM, UN, WB, WTO

💱 Central African CFA
Franc (XAF)

▦ $7.83 bn

💰 $1,528

🏭 Agriculture,
petroleum

Angola

The Kingdom of Kongo originated in northern Angola in the 14th century and, at its peak, covered much of western central Africa. Another state, the Kingdom of Ndongo, ruled southern Angola. The Portuguese arrived in the late 15th century and established settlements along the coast, primarily to purchase slaves to labour in Brazil. Their control over the region caused the indigenous kingdoms to decline. Ndongo collapsed after Portugal pushed inland during the 17th century, while Kongo's territory contracted until it was formally abolished in 1914.

With the local population forced to work on coffee and cotton plantations, rising nationalist sentiment erupted in a guerrilla war against Portuguese colonial forces from 1961–74. Angola won independence in 1975, but was soon torn apart by a civil war between rival political factions. Peace came in 2002, after 1.5 million people had died and 4 million had been displaced. Violence and instability have seriously hampered development, and few Angolans have benefitted from their country's extensive oil reserves.

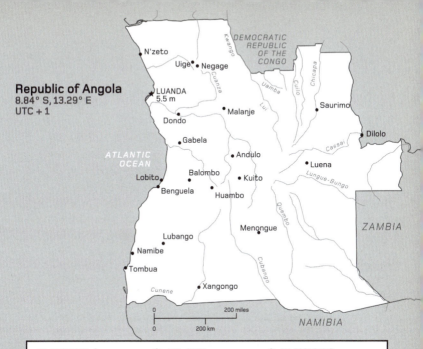

Republic of Angola
8.84° S, 13.29° E
UTC + 1

DEMOCRATIC
REPUBLIC
OF THE
CONGO

Kwango

N'zeto

Uige • Negage

Cuanza

Uamba

Cuilo

Chicapa

LUANDA
5.5 m

• Saurimo

Dondo

• Malanje

Lui

• Dilolo

ATLANTIC
OCEAN

• Gabela

• Andulo

Cassai

• Luena

Lobito

Balombo

• Kuito

Lungue-Bungo

Benguela

Huambo

Lubango

• Menongue

Quembo

ZAMBIA

Namibe

• Tombua

• Xangongo

Cubango

Cunene

NAMIBIA

0		200 miles
0		200 km

☐ 1,246,700 km² (481,354 sq mi)	⬠ 44.8 : 55.2%	❂ AU, ECCAS, IMF, NAM, OPEC, SADC, UN, WB, WTO
O Semi-arid in south and along coast, subtropical elsewhere	ᚲ 37% Ovimbundu 25% Kimbundu 13% Bakongo 25% others	
⸸ 28,813,463		▣ Angolan Kwanza (AOA)
⊞ 23.1/km² (60/sq mi)	⨡ Portuguese	▦ $89.63 bn
↟↟ 3.4%	▢ Christianity	▤ $3,111
	♟ Unitary presidential constitutional republic	⛟ Oil

Namibia

A dry land of desert and savannah in southern Africa, Namibia is the continent's least densely populated country. Its first inhabitants were the Bushmen, nomadic hunter-gatherers who were joined by several other groups, including Bantu tribes, from the 14th century. Germany claimed the territory in 1884, discovering mineral deposits and confiscating land for European settlers. The colonial regime in German South West Africa was brutal; from 1904–7 the Herero and Nama people launched a rebellion. Germany quashed the uprising through systematic genocidal persecution of these groups, killing over four-fifths of the Herero and half of the Nama populations. In 1915, South African forces occupied the colony, and although they did not formally annexe it, they established apartheid rule that lasted for 75 years. From 1966, they faced a guerrilla war with nationalists that lasted until 1990, when South African forces withdrew and the independent Republic of Namibia was established. Since that time, the country has become a multiparty democracy and has actively moved towards national reconciliation.

ZAMBIA

ANGOLA

Cunene

Okavango

Oshakati • Ondangwa

Tsumeb
Otavi • • Grootfontein
Outjo •
• Otjiwarongo

Omaruru •

Usakos • Okahandja •
Gobabis •

Swakopmund • ★ WINDHOEK
368 k

Walvis Bay • • Rehoboth

BOTSWANA

• Maltahohe

Keetmanshoop •

Luderitz •

Grunau •

Karasburg •

ATLANTIC
OCEAN

SOUTH
AFRICA

Orange

Oranjemund •

Republic of Namibia
22.56° S, 17.06° E
UTC + 2

- ☐ 824,290 km²
  (318,260 sq mi)
- ○ Desert
- ♀ 2,479,713
- ☷ 3/km²
  (8/sq mi)
- ♀♀ 2.2%
- ☐ 47.6 : 52.4%
- ♙ 50% Ovambo
  50% others
- ♟ English
- ☐ Christianity
- ☷ Unitary
  semi-presidential
  republic
- ☉ AU, CON, G77, IMF,
  NAM, SADC, UN, WB,
  WTO
- ☐ Namibian Dollar
  (NAD)
- ▨ $10.27 bn
- ▨ $4,140
- ☰ Mining (particularly
  uranium, diamonds
  and zinc)

0   200 miles

0   200 km

Democratic Republic of the Congo

The largest country in sub-Saharan Africa, the DRC originated as a vast tract of land granted to King Leopold II of Belgium, in 1885. This created the Congo Free State, which was essentially a vast rubber plantation overseen by a brutal military force. The country remained Leopold's private property until 1908, when the Belgian government took over. Following independence in 1960, the country descended into crisis, with southern regions threatening to secede and rival political factions battling for control. In 1965, Mobutu Sese Seko seized power, renaming the country Zaire. His corrupt and dictatorial rule saw opposition parties banned while he and his allies enriched themselves from the country's mineral resources, particularly copper, cobalt, gold and diamonds. Mobutu ruled until 1997, when he fled following a rebel victory in the First Congo War. The country, renamed the DRC, descended into war again in 1998. This Second Congo War drew in eight other states and led to the deaths of six million people. Peace was made in 2003, and multi-party elections were held in 2006; however, rebel violence continues to be endemic.

Democratic Republic of the Congo (DRC)
4.44° S, 15.27° E
UTC +1 to UTC +2

CENTRAL AFRICAN REPUBLIC

SOUTH SUDAN

Bondo

Gemena
Buta
Isirio

Lisala
Bumba

Bunia

Basankusu
Isangi
Kisangani

Boende

Mbandaka

Goma

Inongo
Bukavu

Kindu-port-empain
Uvira

Bandundu

Kasai

KINSHASA
11.6 m

Tshela
Madimba

Lusambo
Lomami

Mbanza-Ngungu
Kenge
Kikwit
Luebo

Kalemi

Kananga
Mbuji-mayi
Kabalo

Tshilenge

Congo

Lake Tanganyika

ATLANTIC OCEAN

ANGOLA

Kamina

Pweto

Dilolo
Kolwezi
Likasi

Lubumbashi

ZAMBIA

REP. OF CONGO

UGANDA
RWANDA
BURUNDI
TANZANIA

▢ 2,344,860 km² (905,356 sq mi)	♨ Unitary semi-presidential republic
◯ Tropical	⊕ AU, ECCAS, G24, G77, IMF, NAM, SADC, UN, WB, WTO
⚧ 78,736,153	
⊞ 34.7/km² (90/sq mi)	💷 Congolese Franc (CDF)
⚧ 3.3%	🏦 $34.99 bn
⬠ 43 : 57%	🏦 $445
⚧ 80% Bantu peoples 20% others	⛏ Mining (particularly gold, cobalt and copper), mineral processing
☻ French	
🕮 Christianity	

0   250 miles

0   250 km

Chad

Covered by the Sahara Desert in the north and savannah in the south, Chad is bisected by the Sahel, a belt of semiarid territory. In the centre and north, Muslim states rose from the tenth century; largely nomadic, they were often involved in the trans-Saharan slave trade. Southern areas tended to practise sedentary agriculture, and largely adopted Christianity. In 1900, France established colonial rule over the country. Chad won independence in 1960 and François Tombalbaye, a southerner, became its first president. His regime discriminated against other regions, and following a revolt from 1969–70, he was overthrown in a coup, with a military junta taking power. The country descended into civil war between 1979 and 1982, at the same time facing repeated interventions from Libya that continued until 1987. The government has since attempted to build democracy and reconcile Chad's ethnic, political and religious groups, while developing the country through profits from oil – an export since 2003. However, the ruling regime is increasingly dictatorial, internal violence continues and most of the population still lives in poverty.

Republic of Chad
12.13° N, 15.06° E
UTC + 1

☐	1,284,000 km² (495,755 sq mi)
○	Tropical to desert
✝	14,452,543
⊞	11.5/km² (30/sq mi)
⇈	3.1%
◠	22.6 : 77.4%
⋔	30% Sara 10% Kanembu/ Bornu/Buduma 10% Arab 60% others
⚲	French, Arabic
⌼	Islam, Christianity
⛏	Unitary dominant-party presidential republic
⊕	AU, ECCAS, G77, IMF, NAM, UN, WB, WTO
⌑	Central African CFA Franc (XAF)
▦	$9.60 bn
▦	$664
⛴	Agriculture, oil

LIBYA

0   200 miles

0   200 km

NIGER

• Faya-Largeau

Lake Chad

• Mao

• Biltine

• Bol

• Abeche

SUDAN

• Ati

N'DJAMENA 1.3 m

• Mongo

NIGERIA

• Massenya

• Am Timan

Chari

Bahr Aouk

• Bongor

Logone

• Lai

• Sarh

CENTRAL AFRICAN REPUBLIC

• Moundou

CAMEROON

Central African Republic

The savannahs and forests that make up the CAR were largely isolated until the 17th century, when slave traders ventured into the region in search of captives. During the late 19th century, the area was incorporated into the French empire, which established the colony of Ubangi-Shari (named after the region's two main rivers) in 1903. In 1960, the colony won independence as the CAR, but spent three decades under dictatorial military rule. This included a brief 'imperial' period from 1976–79, when the dictator Jean-Bédel Bokassa proclaimed himself Emperor of Central Africa, before being overthrown with the assistance of French armed forces. Since the 1990s, the country has slowly moved towards multiparty democracy, although it suffered periods of civil war from 2004–7 and 2012–14. According to the UN, the CAR has the lowest level of development in the world, with subsistence agriculture remaining vital, as instability has made it impossible to access the country's potential mineral resources of diamonds, gold and uranium.

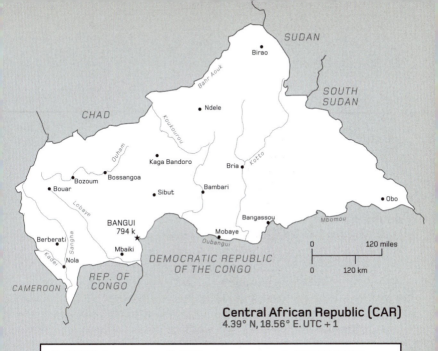

Central African Republic (CAR)
4.39° N, 18.56° E. UTC + 1

☐ 622,980 km² (240,534 sq mi)	♙ 33% Baya 27% Banda 13% Mandjia 10% Sara 17% others	⬙ AU, ECCAS, G77, IMF, NAM, UN, WB, WTO
○ Tropical		▣ Central African CFA Franc (XAF)
✝ 4,594,621	♟ French, Sango	▦ $1.76 bn
▣ 7.4/km² (19/sq mi)	▢ Christianity	▥ $382
↑↑ 1.1%	♟ Semi-presidential republic	▨ Agriculture, forestry
☐ 40.3 : 59.7%		

South Africa

A multi-ethnic state, South Africa had been settled by Bantu peoples by the sixth century; today, they make up three-quarters of the population. In 1652, the Netherlands established the Cape Colony, which Britain seized in 1806. Many Dutch settlers trekked north to establish independent 'Boer Republics'. The discovery of gold and diamonds there during the late 19th century led to war between the British and the Boer. Victorious by 1902, the British incorporated the Boer Republics into South Africa. The country won self-governance in 1931, after which the white ruling minority imposed an apartheid system that separated the population along racial lines and lasted until 1991. Resistance groups fought this, and South Africa became increasingly isolated internationally because of its racist policies. The first multiracial elections were held in 1994, and led to the election of Nelson Mandela as president. Although South Africa has industrialized, poverty and economic inequality still persist, and the country suffers from one of the most serious HIV/AIDS epidemics in the world, with over one-tenth of the population infected.

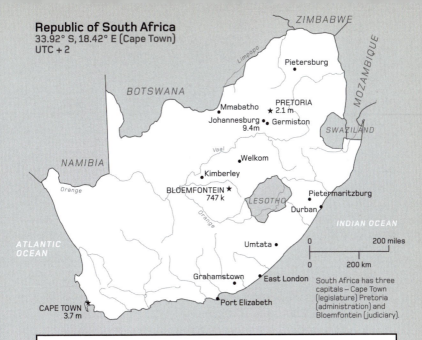

Republic of South Africa
33.92° S, 18.42° E (Cape Town)
UTC + 2

ZIMBABWE

MOZAMBIQUE

Limpopo

BOTSWANA

• Pietersburg

Mmabatho •

PRETORIA ★
2.1 m

Johannesburg • • Germiston
9.4m

SWAZILAND

Vaal

• Welkom

NAMIBIA

Orange

• Kimberley

BLOEMFONTEIN ★
747 k

LESOTHO

Pietermaritzburg •

• Durban

Orange

INDIAN OCEAN

ATLANTIC
OCEAN

Umtata •

0   200 miles

0   200 km

Grahamstown •
• East London

South Africa has three
capitals – Cape Town
(legislature) Pretoria
(administration) and
Bloemfontein (judiciary).

• Port Elizabeth

CAPE TOWN
3.7 m

☐ 1,219,090 km² (470,693 sq mi)	� 80% black African 20% others	⊕ AU, CON, G24, G77, IMF, NAM, SADC, UN, WB, WTO
○ Mostly semi-arid, subtropical along east coast	♟ 11 official languages, most widely spoken is Zulu	▨ South African Rand (ZAR)
⚤ 55,908,865	▯ Christianity	▩ $294.84 bn
▣ 46.1/km² (119/sq mi)	♣ Unitary parliamentary constitutional republic	▤ $5,274
⇅ 1.6%		⚒ Mining (particularly platinum), automobile assembly, metalworking
◔ 65.3 : 34.7%		

Botswana

Two-thirds of Botswana is covered by the Kalahari Desert. The majority of its population originated from Bantu migrants, who arrived from the north by the mid-fifth century. They formed the majority Tswana group, who by the 19th century had established eight principal ruling chiefdoms. Around this time Europeans began to arrive, spreading Christianity and trading for goods such as ivory. Britain established the Bechuanaland Protectorate over the area in 1885, with local rulers remaining influential.

In 1966, the country became independent as Botswana. Since that time there has been an uninterrupted period of civilian rule, and although the Botswana Democratic Party has dominated politics, there are regular free and fair elections. Traditionally, the economy is based on farming and herding, although diamond-mining began in the 1970s. This transformed Botswana, leading to rapid economic growth. Thanks to low levels of corruption and a stable government, profits are reinvested into the country, giving it the highest levels of development in sub-Saharan Africa.

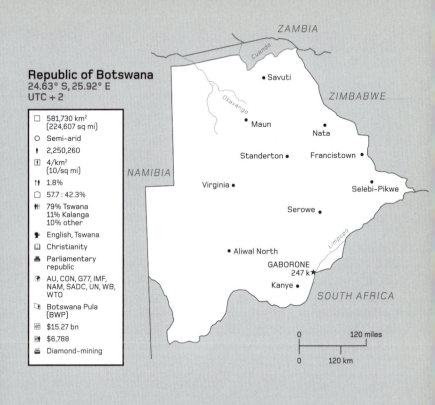

Republic of Botswana
24.63° S, 25.92° E
UTC + 2

☐	581,730 km² (224,607 sq mi)
○	Semi-arid
♀	2,250,260
⊡	4/km² (10/sq mi)
♈	1.8%
◔	57.7 : 42.3%
⚲	79% Tswana 11% Kalanga 10% other
☝	English, Tswana
▢	Christianity
▥	Parliamentary republic
◉	AU, CON, G77, IMF, NAM, SADC, UN, WB, WTO
▧	Botswana Pula (BWP)
▩	$15.27 bn
▦	$6,788
▤	Diamond-mining

ZAMBIA

Cuando

ZIMBABWE

Okavango

• Savuti

• Maun

Nata •

NAMIBIA

Standerton •   Francistown •

• Virginia

Selebi-Pikwe •

Serowe •

Limpopo

• Aliwal North

GABORONE
247 k ★

• Kanye   SOUTH AFRICA

0   120 miles

0   120 km

Sudan

The site of the ancient Nubian civilization that arose in northeastern Africa, Sudanese territory was subject to periodic Egyptian dominance from the fourth millennium BCE until the 11th century BCE, when the Kingdom of Kush was established. It collapsed in 350 CE, after which its territory split into three states. Muslim Arabs migrated to the area from the seventh century onwards, their religion and language becoming adopted widely (they named the region Sudan, which means 'Land of the Blacks'). Sudan came under Egyptian rule in 1821, and from 1899 was effectively governed as a British colony; it won independence in 1956. Since that time the country has been beset by numerous coups, authoritarian leaders and periods of military rule. Internal conflict has been a constant, particularly in the southern, mainly Christian, area of the country, which became independent in 2011. (Now South Sudan, this area contains three-quarters of the region's oil, so its loss seriously damaged the Sudanese economy.) Since 2003, another rebellion in the western Darfur region has involved the deaths of over 200,000 people and displaced two million.

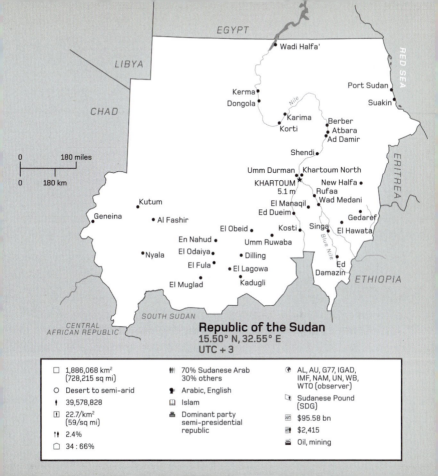

Republic of the Sudan
15.50° N, 32.55° E
UTC + 3

☐ 1,886,068 km² (728,215 sq mi)	�"fi 70% Sudanese Arab 30% others	◉ AL, AU, G77, IGAD, IMF, NAM, UN, WB, WTO (observer)
○ Desert to semi-arid	✶ Arabic, English	⬚ Sudanese Pound (SDG)
✝ 39,578,828	⬚ Islam	▦ $95.58 bn
⊞ 22.7/km² (59/sq mi)	⚒ Dominant party semi-presidential republic	▤ $2,415
✝✝ 2.4%		⬛ Oil, mining
☐ 34 : 66%		

Zambia

Bantu peoples settled Zambian territory in southern Africa from around 300 CE, establishing chieftainships and kingdoms. The area was unexplored by Europeans until a Portuguese expedition in 1798; after centuries of isolation, it then became the focus of Africans and Europeans eager to acquire slaves and ivory. From the late 19th century, the British were the dominant imperial power, establishing the colony of Northern Rhodesia. After winning independence in 1964, the country took the name Zambia (after the Zambezi River). Export incomes were so reliant on copper that, when the price of the mineral dropped in the 1970s, it severely damaged the Zambian economy, forcing the government to borrow heavily. At the same time, the government became authoritarian, banning opposition parties in 1972. Popular protests brought the return of free and fair elections in 1991. The country is now politically stable, with renewed economic growth, thanks to a rise in demand for copper. Such dependence on one mineral makes Zambia highly vulnerable to fluctuations in prices, however, threatening its long-term development.

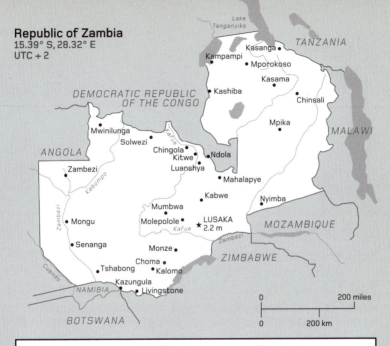

Republic of Zambia
15.39° S, 28.32° E
UTC + 2

Lake Tanganyika

TANZANIA

Kasanga
Kampampi
Mporokoso
Kasama
Kashiba
Chinsali

DEMOCRATIC REPUBLIC
OF THE CONGO

Mpika

MALAWI

Mwinilunga
Solwezi
Chingola
Kitwe
Ndola
Luanshya
Mahalapye

ANGOLA

Zambezi

Kabwe

Nyimba

Mumbwa
Molepolole
LUSAKA
2.2 m

MOZAMBIQUE

Mongu

Senanga

Monze

ZIMBABWE

Choma
Tshabong
Kalomo
Kazungula
Livingstone

NAMIBIA

BOTSWANA

Kafue
Kafue
Zambezi
Kabompo
Cuando

| 0 | | 200 miles |
| 0 | | 200 km |

□ 752,610 km²
(290,584 sq mi)

○ Tropical

16,591,390

22.3/km²
(58/sq mi)

3.0%

41.4 : 58.6%

21% Bemba
14% Tonga
65% others

English, Bemba

Christianity

Unitary presidential
republic

AU, CON, G77, IMF,
NAM, SADC, UN,
WB, WTO

Zambian Kwacha
(ZMW)

$19.55 bn

$1,178

Mining (particularly
copper)

South Sudan

South Sudan's population comprises a range of sub-Saharan peoples, who remained independent until the 19th century when Egypt advanced into the area having conquered northern Sudan. When all of Sudan came under Anglo-Egyptian rule in 1899, the southern area was largely administered separately. Christian missionaries converted the majority of the population from traditional animist beliefs. In 1946, northern and southern Sudan merged, leading to a separatist rebellion that began in 1955 – the year before Sudanese independence. This escalated into a guerrilla war with the Sudanese national government, which lasted until 1972 and ended with the south becoming an autonomous region. The withdrawal of this status in 1983 led to renewed conflict and fighting continued until 2005. South Sudan became independent six years later, following a national referendum. The state has struggled to find stability, with an ongoing civil war between rival political factions starting in 2013. This conflict has displaced over two million people and prevented the country from profiting fully from its resources of oil.

Republic of South Sudan
4.85° N, 31.57° E
UTC + 3

☐ 644,330 km² (248,777 sq mi)	◻ 19 : 81%	☆ AU, EAC, G77, IGAD, IMF, UN, WB
○ Hot with seasonal rainfall	👫 36% Dinka 16% Nuer 48% others	🖼 South Sudanese Pound (SSP)
⚊ 12,230,730	👤 English	📊 $9.01 bn (2015)
🗺 19.7/km² (51/sq mi)	🏛 Christianity, traditional African religions	📈 $759 (2015)
👫 2.9%	♟ Federal presidential constitutional republic	⚒ Oil

Egypt

Due to Egypt's desert landscape, its population has always been concentrated along the Nile River and coastal areas. First unified around 3150 BCE, the kingdom became one of the most important ancient empires under local dynasties, who ruled in three major phases known as the Old, Middle and New Kingdoms, until its conquest by Persian invaders in 343 BCE. Foreign rule then continued under the Greeks, Romans, Byzantines, Arabs and Ottomans. In 1867, it gained autonomy within the Ottoman Empire.

The opening of the Suez Canal in 1869 gave Egypt great strategic importance; Britain established a protectorate over the country in 1882 to ensure access. Full sovereignty was not restored until 1952, and the government nationalized the canal four years later. Although largely politically stable, an authoritarian regime suppressed political opponents. In 2011, uprisings saw the president overthrown, free elections held and Islamists gaining power. Following popular protests in 2013, this government was also overthrown, and the country has since stabilized under more moderate leadership.

MEDITERRANEAN SEA

Alexandria
Port Said
Damanhur
Al Mansurah
CAIRO
18.8 m
El Giza
As Suways
Al Minya
Beni Suef
Al Minya
Mallawi
Asyut
Sohag
Qena
Girga
Luxor
Aswan

Lake
Nasser

ISRAEL

JORDAN

Elat

SAUDI
ARABIA

RED SEA

LIBYA

SUDAN

Nile

0   90 miles
0   90 km

Arab Republic of Egypt
30.04° N, 31.24° E. UTC + 2

☐ 1,001,450 km² (386,662 sq mi)	♛ 99.6% Egyptian 0.4% others	✈ AL, AU, G-15, G24, G77, IM, NAM, UN, WB, WTO
O Mainly desert	✿ Arabic	✎ Egyptian Pound (EGP)
✝ 95,688,681	☐ Islam	▦ $336.30 bn
⊞ 96.1/km² (249/sq mi)	⚒ Unitary semi-presidential republic	▤ $3,514
⇈ 2.0%		⚒ Oil, cotton
☐ 43.2 : 56.8%		

Zimbabwe

Situated between the Zambezi and Limpopo rivers, Zimbabwe's main geographical feature is the Highveld, a plateau that covers the centre of the country. The region was ruled by a succession of Bantu tribes until colonial rule began during the 1890s. The British South Africa Company extended into the area, naming it Southern Rhodesia in honour of its founder, Cecil Rhodes. Declaring independence in 1965 and dropping 'southern' from its name, it was ruled by its white minority until 1980, when international sanctions and a guerrilla war with black nationalists brought free elections. The country became Zimbabwe, named after the Bantu kingdom that had ruled from the 13th to 15th centuries. Robert Mugabe, one of the main nationalist leaders, won the elections, establishing a dictatorial regime, which embarked on land reforms that seized white-owned farms, often violently, and gave them to party connections with little agricultural experience. This contributed to economic decline, which has led to food shortages, inflation and unemployment. Mugabe was forced out of power in 2017, following pressure from the armed forces and his own party.

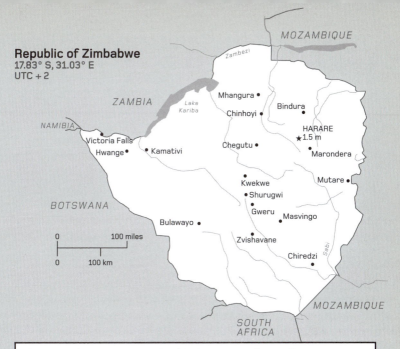

Republic of Zimbabwe
17.83° S, 31.03° E
UTC + 2

MOZAMBIQUE

Zambezi

ZAMBIA

Lake
Kariba

NAMIBIA

Mhangura •

Bindura •

Chinhoyi •

HARARE
★ 1.5 m

Victoria Falls •
Hwange • • Kamativi

Chegutu •

Marondera •

Mutare •

BOTSWANA

Kwekwe •
• Shurugwi
Gweru •

Masvingo •

| 0 | 100 miles |

Bulawayo •

Zvishavane •

| 0 | 100 km |

Chiredzi •

Sabi

MOZAMBIQUE

SOUTH
AFRICA

☐ 390,760 km² (150,873 sq mi)	⚘ Shona (15 other official languages)	▢ United States Dollar (USD)
○ Tropical	▢ Christianity	▦ $16.29 bn
⊡ 16,150,362	⚒ Unitary dominant-party presidential republic	▦ $1,009
⸸ 41.7/km² (108/sq mi)		⚒ Mining (particularly gold), agriculture (particularly tobacco)
⸸⸸ 2.3%	⊕ AU, G-15, G77, IMF, NAM, SADC, UN, WB, WTO	
☐ 32.3 : 67.7%		
⸸⸸ 71% Shona 16% Ndebele 11% others		

Lesotho

The only nation in the world whose entire territory is over 1,000 metres (3,280 ft) above sea level, Lesotho is a mountainous country completely surrounded by South Africa. The Sotho, a Bantu people, began migrating to the region from the 16th century, establishing villages and chiefdoms that coalesced into the Basotho nation under the leadership of a local chief, Moshoeshoe, in 1822. During the mid-19th century, conflict arose with the Boers, who wanted to settle the area and to maintain internal sovereignty. Moshoeshoe appealed to Britain for protection, leading to the creation of the British Protectorate of Basutoland in 1868; following independence in 1966, this became the Kingdom of Lesotho. Moshoeshoe's descendants reigned as constitutional monarchs, thus limiting their formal power. There has been periodic violence between political factions and, from 1986 to 1993, the country was under military rule. Since then, democracy has been re-established, although the country remains poor and is strongly reliant on South Africa, both as a trading partner and as a source of remittances from citizens who work there.

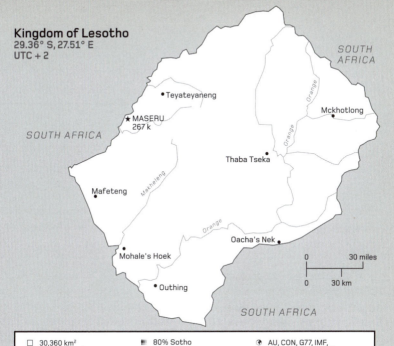

Kingdom of Lesotho
29.36° S, 27.51° E
UTC + 2

SOUTH AFRICA

SOUTH AFRICA

SOUTH AFRICA

• Teyateyaneng

★ MASERU
267 k

Mafeteng

Mohale's Hoek

• Outhing

Thaba Tseka

Oacha's Nek

Mckhotlong

Orange

Orange

Maheleng

Orange

| 0 | | 30 miles |
| 0 | | 30 km |

☐ 30,360 km² (11,722 sq mi)	♟ 80% Sotho 15% Zulu 5% others	⌾ AU, CON, G77, IMF, NAM, SADC, UN, WB, WTO
◯ Temperate	♟ Sotho, English	▱ Lesotho Loti (LSL)
♟ 2,203,821	▥ Christianity	▨ $2.20 bn
▥ 72.6/km² (188/sq mi)	♟ Unitary parliamentary system under constitutional monarchy	▨ $998
♟♟ 1.3%		▰ Agriculture, textiles
◻ 27.8 : 72.2%		

Rwanda

A land of numerous lakes with savannah to the east and mountains to the west, Rwanda is the most densely populated country in mainland Africa. The country originated as a kingdom established in the 18th century by the Tutsi. It became part of German East Africa in 1885, before being placed under Belgian administration after the First World War. In 1959, violence erupted between the Hutu majority and the Tutsi minority, which led to the king being overthrown and the flight of 150,000 Tutsi abroad. Independence was granted in 1962, and the country fell under military rule in 1973. In 1990, Tutsi exiles invaded from Uganda, beginning the Rwandan Civil War, which continued until 1994. Towards the end of the conflict, the Hutu-led government began a mass slaughter of the Tutsi population, killing at least 800,000 people. The Tutsi were ultimately victorious, and established a government of national unity, which has been successful in rebuilding prosperity and stability. Women play a major role in political life; the 2008 elections led to the first parliament in the world with a majority of female members.

Republic of Rwanda
1.97° S, 30.10° E
UTC + 2

DEMOCRATIC
REPUBLIC
OF THE CONGO

UGANDA

TANZANIA

Kakitumba

Akagera

Ruhengeri

Gabiro

Nyabarongo

★ KIGALI
1.3 m

Lake Kivu

Kibuye

Mwongo

Rusizi

TANZANIA

BURUNDI

0   30 miles

0   30 km

☐ 26,340 km² (10,170 sq mi)	👫 85% Hutu 14% Tutsi 1% Twa	🌐 AU, CON, EAC, G77, IMF, NAM, UN, WB, WTO
○ Temperate	👤 Kinyarwanda, English, French, Swahili	💷 Rwandan Franc (RWF)
👤 11,917,508	🏛 Christianity	💰 $8.38 bn
📏 483.1/km² (1,251/sq mi)	♟ Unitary semi-presidential republic	📈 $703
👥 2.4%		🏭 Agriculture, mining (particularly tin)
☐ 29.8 : 70.2%		

Burundi

Although first populated by the pygmy Twa, Burundi's population descends primarily from the Hutu, a Bantu people who settled the area from the 11th century. They were later joined by the Tutsi, who by the late 16th century had established monarchical rule over the country. Burundi became a German protectorate during the late 19th century and was placed under Belgian administration after the First World War.

Burundi won independence as a constitutional monarchy in 1962, but the king was deposed in a coup four years later and a republic declared. Three decades of Tutsi-led military dictatorship followed, with near-constant violence between the two main ethnic groups. A Hutu president was elected in democratic elections in 1993, but Tutsi rebels assassinated him, plunging the country into a bitter civil war that resulted in the deaths of 200,000 people. Peace came in 2005, with a new democratic, power-sharing constitution. Burundi remains underdeveloped and poverty-stricken, with the lowest GDP per capita in the world.

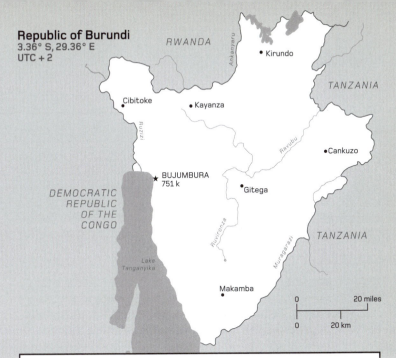

Republic of Burundi
3.36° S, 29.36° E
UTC + 2

RWANDA

TANZANIA

• Kirundo

• Cibitoke

• Kayanza

Ankanyaru

Ravubu

• Cankuzo

★ BUJUMBURA
751 k

Ruzizi

• Gitega

*DEMOCRATIC
REPUBLIC
OF THE
CONGO*

Ruvironza

Muragarazi

TANZANIA

*Lake
Tanganyika*

• Makamba

| 0 | | 20 miles |
| 0 | | 20 km |

☐ 27,830 km² (10,745 sq mi)	ⵜ 81% Hutu 16% Tutsi 3% others	⊕ AU, EAC, ECCAS, G77, IMF, NAM, UN, WB, WTO
○ Equatorial	♟ French, Kirundi	⛁ Burundian Franc (BIF)
⸙ 10,524,117	▣ Christianity	▤ $3.01 bn
▣ 409.8/km² (1,061/sq mi)	⛭ Presidential republic	▦ $286
ⵜⵜ 3.1%		▦ Agriculture
☐ 12.4 : 87.6%		

Tanzania

Bordering three of Africa's great lakes (Victoria, Tanganyika and Nyasa) and containing its highest mountain (Kilimanjaro), Tanzania also includes the Zanzibar archipelago, located in the Indian Ocean. The mainland, known as Tanganyika, became part of Germany's African empire in the late 19th century, before being seized by Britain in 1916. Zanzibar was a base for Arab and Persian traders from the first century CE. After periods of Portuguese and Omani rule, it became an independent sultanate in 1856 that, in 1890, the British declared a protectorate over. Tanganyika became independent in 1961, and a one-party state a year later. After the 1963 Zanzibar Revolution, the sultan was deposed and a communist republic declared. The two territories united in 1964 to form Tanzania, although Zanzibar remained semiautonomous, and continues to have its own domestic government. Despite its composite nature, the new nation was politically stable, and has remained so after the legalization of opposition parties in 1992. However, it remains poor and largely agricultural, although mining and tourism have contributed to recent economic growth.

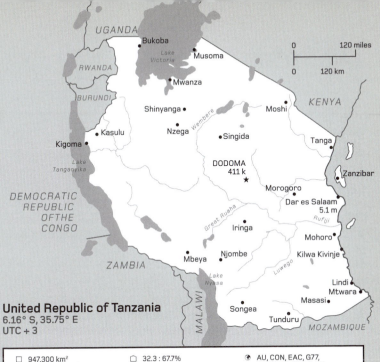

UGANDA

Bukoba •
Lake
Victoria
• Musoma

RWANDA

BURUNDI

• Mwanza

KENYA

Shinyanga •

Moshi •

Wembere

Kasulu •

Nzega •

• Singida

Tanga

Kigoma •

Lake
Tanganyika

DODOMA
411 k
★

Zanzibar

DEMOCRATIC
REPUBLIC
OF THE
CONGO

Morogoro •

Dar es Salaam
5.1 m

Great Ruaha

Rufiji

Iringa •

Mohoro •

ZAMBIA

Mbeya •

Njombe •

Kilwa Kivinje •

Luwego

Lindi •
Mtwara

Lake
Nyasa

Masasi •

United Republic of Tanzania
6.16° S, 35.75° E
UTC + 3

Songea •

Tunduru •

MOZAMBIQUE

MALAWI

☐ 947,300 km² (365,755 sq mi)	☐ 32.3 : 67.7%	☻ AU, CON, EAC, G77, IMF, NAM, SADC, UN, WB, WTO
○ Tropical along coast, temperate in highlands	♚ 95% Bantu 5% others	☖ Tanzanian Shilling (TZS)
✝ 55,572,201	☌ Swahili	☷ $47.43 bn
☷ 62.7/km² (162/sq mi)	☐ Christianity, Islam	☷ $879
↑↑ 3.1%	⚒ Unitary presidential republic	☷ Agriculture, mining (particularly gold, silver and copper)

0   120 miles
0   120 km

Uganda

At the centre of the African Great Lakes region, Uganda has an ethnically diverse population mostly concentrated in the centre and south. Before the 19th century, most of the country was divided into kingdoms; the largest was Buganda, which, in 1894, became the centre of the British Protectorate of Uganda. During colonial rule thousands of Indians were enlisted as indentured labourers to build railway lines; some remained and went on to play a major role in society and commerce. Uganda won independence in 1962, but was divided between ethnic and regional factions, and in 1971 Idi Amin seized power after a military coup. During his eight-year dictatorship, around 300,000 people were killed and the Indian population was expelled. Violence continued after Amin was overthrown in 1979; from 1981–86 Uganda was engulfed in civil war between rival political factions. Since the end of the civil war, Uganda has begun to recover, experiencing economic growth and political stability despite a long-running guerrilla insurgency, led by a rebel group called the Lord's Resistance Army, which only came to an end in 2006.

Republic of Uganda
0.35° N, 32.58° E
UTC + 3

☐	241,550 km² (93,263 sq mi)
○	Mostly tropical
♀	41,487,965
⊡	206.9/km² (536/sq mi)
↟↟	3.3%
◲	16.4 : 83.6%
♯	17% Baganda 10% Banyankole 73% other
♟	English, Swahili
☐	Christianity
♔	Dominant-party semi-presidential republic
◐	AU, CON, EAC, G77, IGAD, IMF, NAM, UN, WB, WTO
◱	Ugandan Shilling (UGX)
⊠	$25.53 bn
◈	$615
⛁	Agriculture (particularly coffee)

SOUTH SUDAN

KENYA

Arua

Nile Albert

Gulu

Achwa

DEMOCRATIC REPUBLIC OF THE CONGO

Pakwach

Lira

Lake Albert

Masindi

Mbale

KAMPALA 1.9 m

Jinja

Kasese

Katonga

Entebbe

Lake Victoria

KENYA

Masaka

Mbarara

Kabale

TANZANIA

RWANDA

0   100 miles

0   100 km

Mozambique

Divided at its centre by the Zambezi River, the northern part of Mozambique is characterized by high plains and mountains, while the south is generally lowland. The area attracted Arab and Persian traders, who, by the 14th century, had settled in coastal areas. They mixed with the indigenous Bantu peoples to create the Swahili culture and language. Portuguese explorers arrived in 1498; after gaining control of the coast they steadily extended their influence inland and, by the 19th century, had established colonial rule over the country. Their oppressive and exploitative regime was met with nationalist resistance, leading to the Mozambican War of Independence, beginning in 1964. Fighting continued for a decade, before Portugal agreed to independence in 1975. The main rebel group established a single-party communist state, but in 1977 the country was plunged into a civil war between the government and opposition forces. Peace was made in 1992, and multi-party elections were held two years later. Since that time political stability has been largely restored, and the country has experienced steady economic growth.

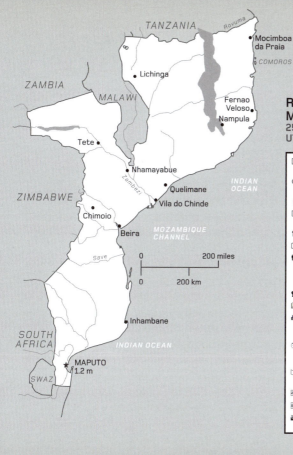

Republic of Mozambique
25.89° S, 32.61° E
UTC + 2

☐	799,380 km² (308,642 sq mi)
○	Tropical to Subtropical
👤	28,829,476
⊞	36.7/km² (95/sq mi)
↑↑	2.9%
◻	32.5 : 67.5%
👥	99.7% African (Makua are the largest group) 0.3% others
👤	Portuguese
📖	Christianity
♣	Unitary semi-presidential republic
⊕	AU, CON, G77, IMF, NAM, SADC, UN, WB, WTO
💱	Mozambican Metical (MZN)
📈	$11.01 bn
💵	$382
🏭	Agriculture

Swaziland

Africa's only absolute monarchy, Swaziland originated as a small state in the southeast of the country, founded around 1770 by settlers from modern-day Mozambique. Conquering neighbouring tribes, it was a powerful regional kingdom by 1860. Encroaching European powers threatened its independence, and in 1894 the country became a protectorate of the Boer Republic of Transvaal. When the Second Boer War ended in 1902, Swaziland transferred to a British administration, resisting attempts to join it to South Africa. Independence was restored in 1968; initially, the country was a democratic constitutional monarchy, but in 1973 the king imposed direct rule. Demand for reform grew in the 1990s, leading to a new democratic constitution in 2005, but royal power remains paramount; political parties are banned, freedom of speech and assembly, restricted and the king has the right to appoint the prime minster and members of parliament. The country is currently in the midst of a health crisis; one-quarter of adults are infected by HIV/AIDS, the highest rate in the world.

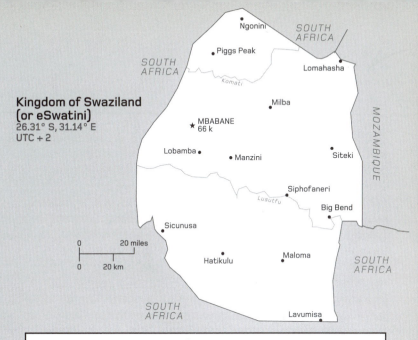

Kingdom of Swaziland
(or eSwatini)
26.31° S, 31.14° E
UTC + 2

Ngonini

SOUTH
AFRICA

Piggs Peak

SOUTH
AFRICA

Lomahasha

Komati

Milba

★ MBABANE
66 k

MOZAMBIQUE

Lobamba

Manzini

Siteki

Siphofaneri

Lusutfu

Big Bend

Sicunusa

0   20 miles

0   20 km

Hatikulu

Maloma

SOUTH
AFRICA

SOUTH
AFRICA

Lavumisa

☐ 17,360 km² (6,703 sq mi)	⚥ 97% African (mostly ethnic Swazi) 3% European	⚙ AU, CON, G77, IMF, NAM, SADC, UN, WB, WTO
○ Tropical to near-temperate	☻ Swazi, English	⛶ Swazi Lilangeni (SZL), pegged to the South African Rand, which is also accepted
✝ 1,343,098	⛪ Christianity	
⊞ 78.1/km² (202/sq mi)	♨ Unitary parliamentary system under absolute monarchy	▥ $3.73 bn
⛁ 1.8%		▤ $2,775
◻ 21.3 : 78.7%		⛏ Agriculture

Malawi

Located along the East African Rift Valley, one-fifth of Malawi's total area is made up of Lake Nyasa. Bantu tribes began to arrive from the fourth century, and in 1480 formed the Maravi Confederacy, which covered most of the country's present territory as well as parts of Zambia and Mozambique. Colonial rule began in the 1890s, with the establishment of the British Central Africa Protectorate, renamed Nyasaland in 1907. Gaining independence in 1964, the country adopted its current name, Malawi. The first prime minister was Hastings Banda, a nationalist leader who proclaimed a republic in 1966. He made himself president, and declared a one-party state, suppressing opponents and killing or imprisoning potential rivals. Following a series of protests, multiparty elections were held in 1994, and Banda was voted out of office. Notwithstanding widespread corruption, democratic rule has been successfully established. However, the Malawian economy remains underdeveloped and heavily dependent on international aid; 80 per cent of the population works in agriculture, which is extremely vulnerable to drought.

Republic of Malawi
13.96° S, 33.77° E
UTC + 2

☐	108,890 km² (42,043 sq mi)
○	Subtropical
✝	18,091,575
▯	191.9/km² (497/sq mi)
✝✝	2.9%
◠	16.5 : 83.5%
👪	35% Chewa 19% Lomwe 13% Yao 12% Ngoni 15% others
👤	English, Chichewa
📖	Christianity
🏛	Unitary presidential republic
🌐	AU, CON, G77, IMF, NAM, SADC, UN, WB, WTO
💵	Malawian Kwacha (MWK)
🔲	$5.44 bn
🔲	$301
🏺	Agriculture

TANZANIA

Chisenga

Rumphi

Nkhata Bay

ZAMBIA

Lake Nyasa

MOZAMBIQUE

Nkhotakota

Kasungu

Mchinji

Salima

★ LILONGWE
905 k

Dedza

Mangochi

Ntcheu

Liwonde

Zomba

MOZAMBIQUE

Mwanza

Blantyre

Ngabu

Nsanje

0   120 miles

0   120 km

Ethiopia

A geographically diverse country featuring mountains, deserts and tropical forests, Ethiopia is amongst the oldest continuously self-governing countries in the world. It is also home to some of the earliest Christian churches, dating to the fourth century CE.

The Ethiopian empire was established in the north of the country by the Zagwe dynasty around 1137. In 1270, it was overthrown by the Solomonic dynasty, which extended its rule over all of modern-day Ethiopia and Eritrea. European colonialism was resisted until 1936, when Italy invaded and annexed the empire. Italian rule was short-lived, however; in 1941, the Allies and Ethiopian resistance defeated Italy, restoring independence. Following famine and protests, a Soviet-backed group – the Derg – overthrew the emperor in 1974, establishing a communist military dictatorship. This regime saw mass killings, famines and war with Somalia, before being overthrown in 1991. Ethiopia's first-ever democratic elections were held in 1995. Since then the country has been stable, and although it remains poor, its economy has grown steadily over the past decade.

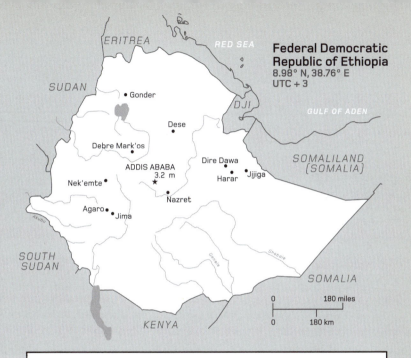

Federal Democratic Republic of Ethiopia
8.98° N, 38.76° E
UTC + 3

ERITREA
RED SEA
SUDAN
GULF OF ADEN
DJI
SOMALILAND
(SOMALIA)

Gonder
Dese
Debre Mark'os
ADDIS ABABA
3.2 m
Dire Dawa
Harar
Jijiga
Nek'emte
Nazret
Agaro
Jima

Akobo
Gibele
Shebele

SOUTH
SUDAN
SOMALIA
KENYA

| 0 | | 180 miles |
| 0 | | 180 km |

☐	1,104,300 km² (426,373 sq mi)	♀♂	34% Oromo 27% Amhara 39% other	☀	AU, G24, G77, IGAD, IMF, NAM, UN, WB, WTO (observer)
○	Tropical monsoon	⚐	Amharic	☐	Ethiopian Birr (ETB)
✝	102,403,196	⚏	Christianity	▦	$72.37 bn
⊡	102.4/km² (265/sq mi)	♨	Federal parliamentary republic	▩	$707
⚦	2.5%			⚒	Agriculture (particularly coffee)
◯	19.9 : 80.1%				

Kenya

The Kenyan coast of East Africa was once the site of many Swahili city-states that established trade links across Asia. In comparison, the plains and highlands of the interior were relatively sparsely populated, by different ethnic groups. British rule began in 1888, and thousands of Asian and European settlers migrated to the colony, now named Kenya after its highest mountain. From 1952–60, the Mau Mau rebels launched an uprising against British rule, winning the country's independence in 1963. The famed nationalist Jomo Kenyatta was elected to lead the country and served as president until he died in 1978. His Kenya African National Union party dominated politics until 2002, when it lost to a coalition of opposition groups. In recent years, Kenya's multi-ethnic population has led to tension and outbreaks of violence; in 2010, a new constitution was promulgated that devolved local power to 47 counties. Although it suffers from drought and unemployment, Kenya has become the main economic hub of East Africa and the wealthiest country in the region, with tea and tourism being particularly important.

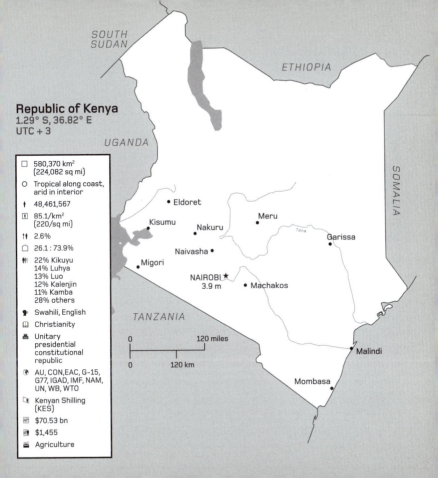

Republic of Kenya
1.29° S, 36.82° E
UTC + 3

SOUTH SUDAN

ETHIOPIA

UGANDA

SOMALIA

☐	580,370 km² (224,082 sq mi)
○	Tropical along coast, arid in interior
👤	48,461,567
⊡	85.1/km² (220/sq mi)
👥	2.6%
▢	26.1 : 73.9%
👫	22% Kikuyu 14% Luhya 13% Luo 12% Kalenjin 11% Kamba 28% others
👤	Swahili, English
🕮	Christianity
🏛	Unitary presidential constitutional republic
🌐	AU, CON,EAC, G-15, G77, IGAD, IMF, NAM, UN, WB, WTO
🏦	Kenyan Shilling (KES)
🏧	$70.53 bn
💵	$1,455
🚜	Agriculture

• Eldoret

Kisumu • Nakuru Meru

Naivasha • Tana Garissa

Migori • NAIROBI ★
3.9 m
• Machakos

TANZANIA

Malindi

Mombasa

0	120 miles
0	120 km

Eritrea

From the second to the tenth centuries, Eritrean territory was part of the Kingdom of Aksum, a great trading power in northeastern Africa. It then fragmented into local sultanates, with the Ottoman empire ruling the coastal area. Italian rule followed in 1890, and lasted until the Italians were expelled by Allied forces during the Second World War. Britain administered the country until 1952, when it became an autonomous part of Ethiopia with its own legislature. A war of independence began in 1961, between Eritrean nationalists and the Ethiopian government. Ending with Ethiopian defeat in 1991, it was followed by a referendum in 1993, in which the Eritrean people overwhelmingly voted for independence. Since then the country has been a one-party state, with an authoritarian government that has used the threat of renewed conflict with Ethiopia to enforce compulsory, indefinite conscription; state repression has prompted thousands to flee the country. Eritrea has been unable to take advantage of its strategic position on the Red Sea, and over 80 per cent of the population is engaged in subsistence farming.

SUDAN

RED SEA

• Keren

Ak'ordat •　　• Massawa

ASMARA ★
804k

• Adi Kwala　• Senafe

ETHIOPIA

RED SEA

DJIBOUTI

0　　　　　　　100 miles

0　　　　　　　100 km

State of Eritrea
15.32° N, 38.93° E
UTC + 3

☐ 101,000 km² (38,996 sq mi)	♟ Tiginya, Arabic
○ Coastal desert, cooler central highlands, semi-arid lowlands	▣ Christianity, Islam
† 4,474,690 (2011)	♙ Unitary one-party presidential republic
⊡ 44.3/km² (115/sq mi)	◉ AU, G77, IGAD, IMF, NAM, UN, WB
†† 1.9% (2011)	▦ Eritrean Nakfa (ERN)
◯ 21 : 79%	▥ $2.60 bn
♛ 55% Tigrinya 30% Tigre 15% others	▨ $583
	⛴ Agriculture

Somalia

The Somali landscape is mostly flat, dry and unproductive, but owing to its strategic location on the Horn of Africa it was once an important trading centre. During the late 19th century, Britain and Italy established colonial regimes here, but Italian rule ended in 1941. After the Second World War, the UN administered the former Italian colony as a trust, and in 1960 it joined with the former British protectorate to form an independent nation. Nine years later, its democratically elected president was assassinated; an authoritarian socialist regime then ruled until 1991, when the country descended into civil war. Central political authority completely collapsed and the country fragmented into numerous local factions, with two of the more stable regions in the north (Somaliland and Puntland) gaining autonomy. In 2004, a transitional government re-established control, instituting a new federal constitution in 2012. Despite gradual progress towards stability, fighting with Islamist insurgents in the south continues, and Somalia is struggling to resolve long-term problems, such as clan conflict, mass internal displacement and rampant corruption.

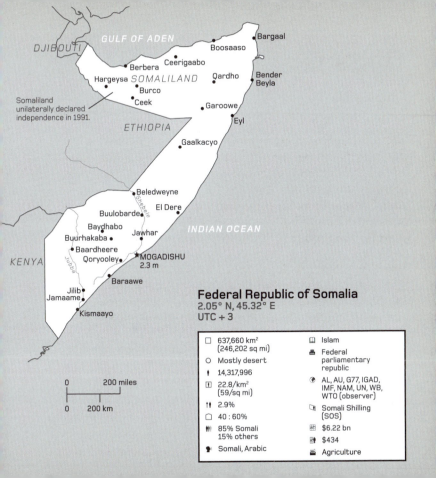

GULF OF ADEN

DJIBOUTI

Bargaal

Boosaaso

Ceerigaabo

Berbera

SOMALILAND

Qardho

Bender Beyla

Hargeysa

Burco

Ceek

Garoowe

Somaliland
unilaterally declared
independence in 1991.

ETHIOPIA

Eyl

Gaalkacyo

Beledweyne

Shebele

El Dere

Buulobarde

Baydhabo

INDIAN OCEAN

Jawhar

Buurhakaba

Baardheere

Qoryooley

★MOGADISHU
2.3 m

KENYA

Jubba

Baraawe

Jilib

Jamaame

Kismaayo

0		200 miles
0		200 km

Federal Republic of Somalia
2.05° N, 45.32° E
UTC + 3

▢	637,660 km² (246,202 sq mi)	📖	Islam
○	Mostly desert	♟	Federal parliamentary republic
♀	14,317,996	☯	AL, AU, G77, IGAD, IMF, NAM, UN, WB, WTO (observer)
▯	22.8/km² (59/sq mi)		
↑↑	2.9%	💷	Somali Shilling (SOS)
▢	40 : 60%	📊	$6.22 bn
♀♀	85% Somali 15% others	📈	$434
♟	Somali, Arabic	⚒	Agriculture

Djibouti

On the southern entrance to the Red Sea, just 32 km (20 miles) from the Arabian Peninsula, Djibouti was one of the first parts of Africa to adopt Islam, and remained an independent sultanate until the 1869 opening of the Suez Canal increased its strategic importance. France, which had already acquired a port in the region, extended its influence to establish the colony of French Somaliland in 1888. Referenda in 1958 and 1967 saw the population vote to remain associated with France, but 1977 saw a landslide victory for independence. The new state was named after Djibouti City, which the French had founded as their colonial capital. The country's first president, Hassan Gouled Aptidon, instituted a repressive one-party state, ruling until 1999. Today, political life is dominated by the Somali Issa clan, which marginalized the minority ethnic group, the Afar, leading to warfare in the northeast of the country that continued from 1991 to 2001. Djibouti has scant natural resources and limited industrial activity; its deep water port in the capital is the long-term foundation of the economy.

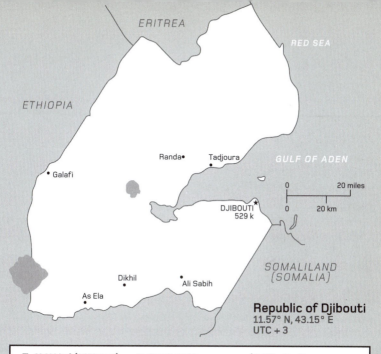

ERITREA

RED SEA

ETHIOPIA

Randa• •Tadjoura

GULF OF ADEN

• Galafi

0   20 miles

0   20 km

DJIBOUTI
529 k

SOMALILAND
(SOMALIA)

• Dikhil   • Ali Sabih

• As Ela

Republic of Djibouti
11.57° N, 43.15° E
UTC + 3

☐ 23,200 km² (8,958 sq mi)	♣ French, Arabic	Djiboutian Franc (DJF)
○ Desert	Islam	$1.73 bn (2015)
942,333	Unitary dominant-party presidential republic	$1,862 (2015)
40.7/km² (105/sq mi)		
1.6%	AL, AU, G77, IGAD, IMF, NAM, UN, WB, WTO	Shipping and port services
77.4 : 22.6%		
60% Somali 35% Afar 5% others		

Comoros

The Comoros archipelago comprises four main islands that lie between Mozambique and Madagascar. Its first inhabitants were Malayo-Indonesians who arrived during the sixth century CE; from around 700, settlers from continental Africa joined them, as well as Arabs and Persians, who introduced Islam. Ruled by local sultans, the islands functioned as an important trading hub, before coming under French colonial rule in the mid-19th century. In 1974, referenda on three of the islands (Grande Comore, Mohéli and Anjouan) voted for independence, which was declared the next year. A fourth island, Mayotte, voted against independence and continues to be under French administration. Instability has been endemic in independent Comoros, with numerous coups contributing to a rapid turnover of regimes. The economy is largely undeveloped and relies heavily on exporting vanilla, cloves and ylang-ylang. Political upheaval saw Mohéli and Anjouan attempting to secede in 1997. To prevent the country breaking apart, however, power-sharing agreements of 2000–1 gave each island domestic autonomy, with the federal presidency rotating between them.

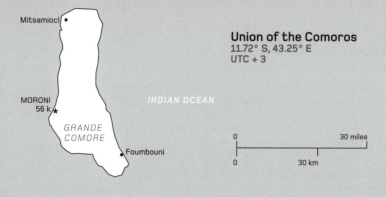

Mitsamiocl •

Union of the Comoros
11.72° S, 43.25° E
UTC + 3

INDIAN OCEAN

MORONI
56 k ★

*GRANDE
COMORE*

Foumbouni •

0		30 miles
0	30 km	

INDIAN OCEAN

Fomboni •

MOHÉLI

Mutsamudu •

ANJOUAN

☐ 1,861 km² (719 sq mi)	♙ 97% Comorian 3% others	🏷 Comorian Franc (KMF)
○ Tropical marine	♟ Comorian, Arabic, French	🖩 $616.65 m
✝ 795,601	📖 Islam	🏛 $776
⊞ 427.5/km² (1107/sq mi)	⚖ Federal presidential republic	🚜 Agriculture
♒ 2.3%		
◻ 28.4 : 71.6%	🌐 AL, AU, G77, IMF, NAM, SADC, UN, WB, WTO (observer)	

Madagascar

Located some 400 km (250 miles) off the African continent, Madagascar's isolation has made it one of the most diverse environments on Earth; more than 80 per cent of its plants and animals are unique to the island. It was uninhabited by humans until settlers from Borneo arrived in the mid-sixth century. Bantu peoples joined them from East Africa after 1000 CE; the two groups combined to form the Malagasy ethnic group. The indigenous Merina Kingdom dominated the island from the mid-16th century until 1883, when France launched a series of campaigns, bringing down the monarchy in 1896. The island then fell under French colonial rule.

With independence restored in 1960, the country endured one-party socialist rule from 1975–92. Protests in 2009 led to the fall of the government, with the army appointing the new president. However, following international mediation, free and fair elections were held in 2013. Madagascar is one of the poorest countries on Earth; although it has potentially lucrative mineral reserves, these are yet to be fully exploited.

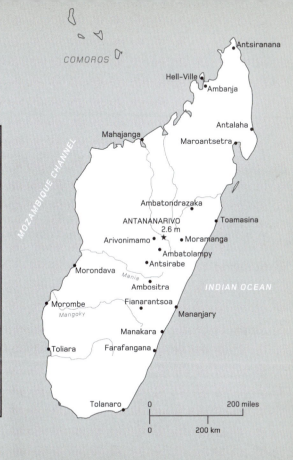

Republic of Madagascar
18.88° S, 47.51° E
UTC + 3

- ☐ 587,295 km²
  (226,756 sq mi)
- ◯ Tropical along
  coast, temperate
  inland, arid in south
- ⚥ 24,894,551
- ⊞ 42.8/km²
  (111/sq mi)
- ⇈ 2.7%
- ◖ 35.7 : 64.3%
- ⚥ 96% Malagasy
  4% others
- ⚑ Malagasy, French
- ⚏ Traditional
  indigenous beliefs,
  Christianity
- ♟ Unitary
  semi-presidential
  constitutional
  republic
- ⚕ AU, G77, IMF, NAM,
  SADC, UN, WB, WTO
- ⎙ Malagasy Ariary
  (MGA)
- ▦ $9.99 bn
- ✂ $401
- ⛴ Agriculture, fishing

COMOROS

MOZAMBIQUE CHANNEL

Antsiranana

Hell-Ville
Ambanja

Antalaha

Mahajanga

Maroantsetra

Ambatondrazaka

ANTANANARIVO
2.6 m

Toamasina

Arivonimamo ★ Moramanga

Ambatolampy

Antsirabe

Morondava

Mania

Ambositra

INDIAN OCEAN

Morombe

Fianarantsoa

Mangoky

Mananjary

Manakara

Manakara

Toliara

Farafangana

Tolanaro

0   200 miles

0   200 km

Seychelles

An archipelago of 115 islands in the western Indian Ocean, the Seychelles were largely uninhabited until French settlement began in the mid-18th century. The islands formally came under British rule in 1814, although French land-owning elites remained in place and the slave-based plantation economy continued. When slavery was abolished in 1835, indentured labourers began to arrive, mostly from India. As such, the vast majority of Seychellois are a mixture of African, Asian and European ancestry. The country gained independence in 1976, but its first elected president was overthrown just one year later. A single-party socialist state was subsequently established in 1979. Surviving several coup attempts, this lasted until a new constitution was promulgated in 1993, when free and fair elections were held; the country has since become a stable multi-party democracy. Shifting its focus from agriculture (particularly coconut, vanilla and cinnamon) to the tourist sector has resulted in a sevenfold increase in per capita GDP – the highest in Africa.

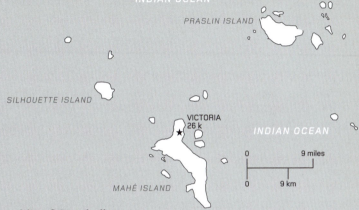

PRASLIN ISLAND

SILHOUETTE ISLAND

VICTORIA
26 k

INDIAN OCEAN

MAHÉ ISLAND

| 0 | | 9 miles |
| 0 | 9 km | |

Republic of Seychelles
4.62° S, 55.45° E
UTC + 4

☐ 460 km² (178 sq mi)	♼ 93% Creole 7% others	⊕ AU, CON, G77, IMF, NAM, SADC, UN, WB, WTO
O Tropical marine	♟ French, English, Seychellois Creole	⃟ Seychellois Rupee (SCR)
⚥ 94,677	☐ Christianity	▦ $1.43 bn
⊞ 206/km² (534/sq mi)	♧ Unitary presidential republic	▦ $15,076
♁ 1.3%		⚒ Tourism, fishing
⬠ 54.2 : 45.8%		

Mauritius

Located in the Indian Ocean, Mauritius is 1,900 km (1,200 miles) off mainland Africa. The island had no permanent human inhabitants until the Dutch established a colony in 1638. They named the main island after their ruler Prince Maurice of Nassau. Being so isolated, Mauritius was the only location of the dodo bird, which was hunted into extinction by 1662. Dutch settlements were abandoned in 1710, and the French took over five years later. They established a naval base and sugar-cane plantations that used sub-Saharan African slaves. Britain conquered Mauritius in 1810; sugar continued to be important, relying on indentured labourers from India after the abolition of slavery in 1835. Mauritian independence was proclaimed in 1968, and in 1992 the island became a republic, with an elected president replacing the British monarch as the (largely ceremonial) head of state. The country is highly stable, with a well-established democracy and regular free elections. The economy has diversified away from sugar, developing tourism and financial services that have made the country one of the most prosperous in Africa.

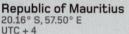

Republic of Mauritius
20.16° S, 57.50° E
UTC + 4

□	2,040 km² (788 sq mi)
○	Tropical
†	1,263,473
⊞	622.4/km² (1,612/sq mi)
††	0.1%
◠	39.5 : 60.5%
†††	68% Indo-Mauritian 25% Creole 7% others
♟	English
◲	Hinduism Christianity
♜	Unitary parliamentary republic
◉	AU, CON, G77, IMF, NAM, SADC, UN, WB, WTO
◱	Mauritian Rupee (MUR)
▦	$12.16 bn
▨	$9,628
⛏	Agriculture, tourism

INDIAN OCEAN

Poudre d'Or

Pamplemousses

★ PORT LOUIS
135 K

Moka

Centre de Flacq

Rose Hill

Quartier Militaire

Tamarin

Nouvelle France

Mahebourg

Souillac

INDIAN OCEAN

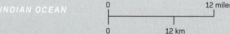

| 0 | | 12 miles |
| 0 | 12 km | |

Asia

The world's largest continent, Asia covers one-third of Earth's landmass and contains nearly two-thirds of its population. Its diversity matches its vast size; Asia contains the highest and lowest points in the world and has a climate that ranges from arctic to tropical. The continent features sparsely peopled areas, such as the Gobi Desert, as well as some of the world's largest cities, including Shanghai, Mumbai and Tokyo.

Many of the world's first civilizations arose in Asia, notably in Mesopotamia, the Indus Valley and northwestern China. The continent was also the base of some of history's largest empires – Persian, Chinese, Arab and Mongol. The world's most practised religions (Christianity, Islam, Hinduism and Buddhism) all originated here. For much of human history Asian nations, such as those of India and China, were the most powerful and advanced in the world, although in the 18th and 19th centuries European powers began to surpass them. Recent Asian development has been rapid and the continent may yet resume its former prominence.

Palestine

Two noncontiguous areas make up Palestine: the Gaza Strip on the Mediterranean coast and the landlocked West Bank. It has its origins in the 1947 UN Partition Plan, which sought to create independent Arab and Jewish states in the region. Owing to the outbreak of warfare between Israel and its Arab neighbours in 1948, the plan was not enacted; Jordan annexed the West Bank and Egypt occupied the Gaza Strip. After the Six-Day War of 1967, Israel won control of both territories. Egypt and Jordan surrendered their territorial claims in 1988, after which Palestine declared independence. It was not until the 1993 First Oslo Accord that Palestinians exercised some measure of self-government. However, Israel retains a high degree of control over Palestinian territory and Israeli settlers remain in the West Bank. Since 2007, the country has been divided between two political factions, with Hamas governing the Gaza Strip and Fatah the West Bank. Palestine remains relatively poor and underdeveloped, with violence, instability and thousands of Palestinians still internally displaced.

State of Palestine

31.90° N, 35.20° E
UTC + 2

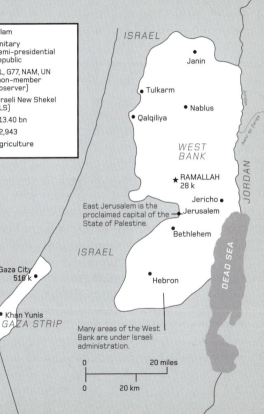

☐	6,220 km² (2,402 sq mi)	🕌	Islam
O	Temperate	♜	Unitary semi-presidential republic
✝	4,551,566	⊕	AL, G77, NAM, UN (non-member observer)
⊡	756.1/km² (6020/sq mi)	💵	Israeli New Shekel (ILS)
⇈	2.9%	💰	$13.40 bn
◖	75.5 : 24.5%	💴	$2,943
🚻	83% Arab 17% others	🌾	Agriculture
🐾	Arabic		

ISRAEL

• Janin

• Tulkarm

• Nablus

• Qalqiliya

WEST BANK

★ RAMALLAH
28 k

• Jericho

East Jerusalem is the
proclaimed capital of the
State of Palestine.

• Jerusalem

• Bethlehem

JORDAN

DEAD SEA

MEDITERRANEAN SEA

ISRAEL

Gaza City
516 k •

• Hebron

• Khan Yunis

GAZA STRIP

Rafah •

EGYPT

Many areas of the West
Bank are under Israeli
administration.

0   20 miles

0   20 km

Israel

Israel's roots lie in the 19th-century Zionist movement that aimed to found a modern Jewish homeland. Zionism's focus was on returning to Palestine, which had been ruled by the Ottoman empire since 1516. The first Jewish settlers arrived from Europe in 1881, and thousands more followed in subsequent decades. Britain administered the area from 1920–48, but there was near-constant violence between British authorities, Jewish settlers and Arab Palestinians. In 1948, Israel declared independence, leading to invasion by Arab states. This was the first of a series of conflicts between Israel and Arab nations, arising from territorial disputes (particularly over Jerusalem) and the status of displaced Palestinians. Israel survived these wars, but there remains considerable tension between Israel and Palestine. Under Israeli law, all Jews worldwide have the right to live there, meaning the country's population has arrived from a wide geographic area. Even though most Israeli territory is arid desert with scarce access to water, it has become one the most prosperous states in the Middle East, with a technologically advanced economy.

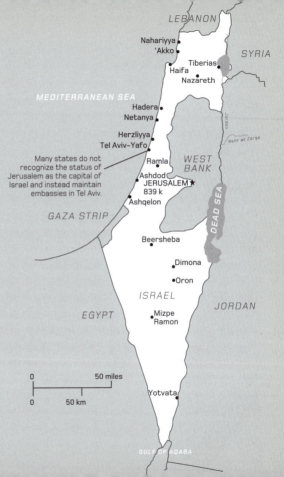

State of Israel
31.78° N, 35.22° E
UTC + 2

☐	22,070 km² (8,521 sq mi)
○	Mostly temperate
✦	8,547,100
⊞	395/km² (1023/sq mi)
⇈	2%
◠	92.2 : 7.8%
⋔	75% Jewish 25% others (mostly Arab)
✦	Hebrew, Arabic
▢	Judaism
♖	Unitary parliamentary republic
⊕	IMF, OECD, UN, WB, WTO
▨	Israeli New Shekel (ILS)
▦	$318.74 bn
▨	$37,293
▤	High-tech products (software and telecommunications), pharmaceuticals

LEBANON

SYRIA

Nahariyya
'Akko
Tiberias
Haifa
Nazareth

MEDITERRANEAN SEA

Hadera
Netanya

Herzliyya
Tel Aviv-Yafo

Ramla

WEST BANK

Many states do not recognize the status of Jerusalem as the capital of Israel and instead maintain embassies in Tel Aviv.

Ashdod
JERUSALEM ★
839 k

DEAD SEA

Ashqelon

GAZA STRIP

Beersheba

Dimona
●Oron

ISRAEL

JORDAN

EGYPT

Mizpe Ramon

0	50 miles
0	50 km

Yotvata

GULF OF AQABA

Saudi Arabia

The Saudi Arabian landscape is mostly dry, arid desert. It was the home of the prophet Muhammad, who founded Islam in the early seventh century. The country houses the two holiest sites in Islam – Mecca and Medina – both destinations for millions of Muslim pilgrims every year. Muhammad united the nomadic Arab tribes. Under his leadership, and that of his successors, the tribes embarked on a series of conquests that, in just over a century, created an Arab empire covering the Middle East, North Africa and most of Iberia. From the 16th century, Saudi Arabia was under Ottoman rule but an Arab uprising overthrew the Turks during the First World War. In 1932, Saudi Arabia became a kingdom ruled by the Saud dynasty; six years later its fortunes were transformed when vast oil reserves were found along its coast. As a result, it became the world's largest oil exporter. The Saud dynasty continues to rule the country as absolute monarchs, although their autocratic regime has attracted criticism, particularly for its gender inequality and support of ultra-conservative Islam.

Ar'ar
Sakakah
Haql
Tabuk
King Khalid
Military City
Ha'il
Al Wajh
Umm
Lujj
Medina
Yanbu
al Bahr
Halaban
Zalim
Mecca
Al-Taif
Jiddah
Qal'at
Bishah
Al
Qunfudhah
Khamis
Mushayt
Jizan
Najran

Buraydah
Rumah
Hafar
al Batin
Ad Dammam
Dhahran
★ RIYADH
6.5 m
As Salwa
As Sulayyil

JORDAN
EGYPT
IRAQ
KUWAIT
PERSIAN
GULF
BAHRAIN
QATAR
UNITED ARAB
EMIRATES
RED
SEA
OMAN
YEMEN

Kingdom of Saudi Arabia
24.71° N, 46.68° E
UTC + 3

☐ 2,149,690 km² (830,000 sq mi)	⚥ 90% Arab 10% Afro-Asian	🏷 Saudi Riyal (SAR)
○ Dry desert	👤 Arabic	💰 $646.44 bn
👤 32,275,687	📖 Islam	💵 $20,029
▦ 15/km² (29/sq mi)	⚖ Absolute monarchy	⛽ Oil and natural gas
👫 2.3%	🌐 ACD, AL, GCC, G77, IMF, NAM, OPEC, UN, WB, WTO	
☐ 83.3 : 16.7%		

0   200 miles

0   200 km

Jordan

Located to the east of the Jordan River, the Transjordan region fell under Roman control from the first century BCE and was subsequently part of the Byzantine empire. A historic watershed occurred in the early seventh century, when Muslim Arabs conquered the region, introducing their religion and language. Ottoman rule began in 1516, continuing for over four centuries, until the Arab Revolt during the First World War. In 1921, a leading figure of the Hashemite clan that had spearheaded the revolt was named Emir of Transjordan. Although the territory enjoyed some autonomy, it was officially a British protectorate and did not become fully independent until 1946, when it was also redesignated a kingdom, taking the name Jordan. Since that time, despite a series of conflicts in the wider region – and an influx of millions of refugees – the country has been mostly stable and a promoter of peace. Following the adoption of a constitution in 1952, there has been a shift towards parliamentary democracy, although the Jordanian king retains a high degree of power.

Sea of
Galilee

SYRIA

IRAQ

Yarmuk

Irbid

Ajtun

Al Mafraq

Jarash

WEST
BANK

As Salt

Ar Ruayshid

★ AMMAN
1.2m

Al Karak

Madaba

Al Umari

Dead
Sea

SAUDI ARABIA

ISRAEL

At Tafilah

Ma'an

SAUDI ARABIA

Al Aqabah

Gulf of
Aqaba

☐	89,320 km² (34,487 sq mi)
○	Mostly arid desert
✝	9,455,802
⊞	106.5/km² (276/sq mi)
††	3.2%
⌂	83.9 : 16.1%
♛	98% Arab 2% others
♟	Arabic
📖	Islam
♖	Unitary parliamentary system under constitutional monarchy
⊕	AL, G77, IMF, NAM, UN, WB, WTO
⌦	Jordanian Dinar (JOD)
▦	$38.65 tn
▨	$4,088
▤	Clothing manufacture, tourism

**The Hashemite
Kingdom of Jordan**
31.95° N, 35.93° E
UTC + 2

0 _____ 50 miles

0 _____ 50 km

Lebanon

A succession of imperial powers, including the Egyptians, Romans, Arabs, Ottomans and French have ruled Lebanon. The country has a long tradition of religious diversity and was an important centre of early Christianity, as well as the home of the Maronite Church, with its own traditions and beliefs. Islam was introduced in the seventh century, but other faiths flourished, such as the Druze, a monotheistic and secretive religion founded in the 11th century that blended Islam with other beliefs. Following independence from France in 1943, a National Pact guaranteed power-sharing between religions. Under this 'confessionalist' system, certain government positions are reserved for particular faiths; for example, the president must be Maronite, the prime minister Sunni, and the deputy prime minister Greek Orthodox. Sectarian tension led to civil war (1975–90); during this conflict the Syrian military occupied the country and did not depart until 2005. The following year, the country was the site of a brief war between Israel and mainly Shia paramilitaries. Since then, refugees from the civil war in neighbouring Syria have added to political tensions.

Lebanese Republic
33.89° N, 35.50° E
UTC + 2

☐	10,450 km² (4,035 sq mi)
○	Mediterranean
✝	6,006,668
⊡	587.2/km² (1522/sq mi)
↑↑	2.6%
◠	87.9 : 12.1%
⁂	95% Arab (inc. 37% Christian Lebanese, many of whom identify as Phoenician) 5% others
✿	Arabic, French
▭	Islam
♨	Unitary parliamentary multi-confessionalist republic
◍	AL, G24, G77, IMF, NAM, UN, WB, WTO (observer)
◫	Lebanese Pound (LBP)
▦	$47.54 tn
▨	$7,914
▥	Banking, tourism, agriculture, jewellery production

MEDITERRANEAN SEA

Qoubaiyat •

Tripoli •

Orontes

• Aynata

Nahr al Litani

BEIRUT
2.3 m ★   • Baabda

Zahle •

Joub Jannine •   SYRIA

Sidon •

Nahr al Hasbani

Nabatiye •

• Tyre

Rmaich •

ISRAEL

0		20 miles
0		20 km

Syria

The geographical variety of Syria, with its Mediterranean coast, interior desert and mountains, is matched by its cultural and religious diversity. The modern state has roots in a French mandate established in 1924, and gained independence as a parliamentary democracy in 1946. The country was politically unstable, and from 1958–61 joined with Egypt as the United Arab Republic. Following a 1963 coup, the Arab nationalist Ba'ath Party won power and has been the ruling political faction since that time. The country has been highly involved in conflict in the region, losing control of the Golan Heights to Israel in 1967 and occupying Lebanon from 1976–2005. As a result of this political turmoil (as well as the impact of drought), recent Syrian economic growth has been slow, despite its oil deposits and fertile plains. The Syrian Civil War has compounded this: breaking out in 2011 following anti-government protests, the conflict has drawn in other countries as well as an array of armed opposition groups, and has created a major humanitarian crisis, resulting in more than 400,000 deaths and displacing millions.

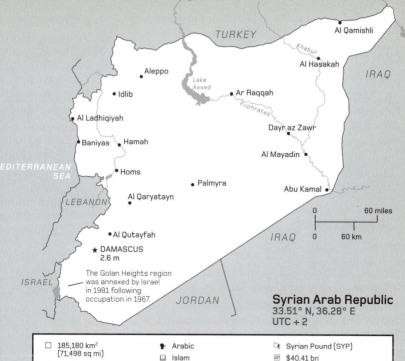

TURKEY

Al Qamishli

Khabur

Aleppo

Al Hasakah

IRAQ

Idlib

Lake
Assad

Ar Raqqah

Al Ladhiqiyah

Euphrates

Dayr az Zawr

Baniyas

Hamah

Al Mayadin

MEDITERRANEAN
SEA

Homs

LEBANON

Palmyra

Abu Kamal

Al Qaryatayn

0   60 miles

IRAQ

Al Qutayfah

0   60 km

★ DAMASCUS
2.6 m

The Golan Heights region
was annexed by Israel
in 1981 following
occupation in 1967.

ISRAEL

JORDAN

Syrian Arab Republic
33.51° N, 36.28° E
UTC + 2

☐	185,180 km² (71,498 sq mi)	✦	Arabic	⬛	Syrian Pound (SYP)
○	Mostly desert	☐	Islam	💰	$40.41 bn (2007)
✝	18,430,453	⚖	Unitary dominant-party semi-presidential republic	📊	$2,058 (2007)
⬓	100.4/km² (260/sq mi)			⚒	Oil, textiles, agriculture
↕	-1.6%	⊕	AL (suspended as of 2017), G24, IMF, NAM, UN, WB, WTO (observer)		
⬒	58.1 : 41.9%				
⬛	90% Arab 10% others				

Iraq

Mesopotamia, situated between the Tigris and Euphrates rivers, was once home to ancient civilizations whose innovations include the first legal codes and writing systems. Much of this region is now located within Iraq, whose modern borders were established in 1920, following the dissolution of the Ottoman Empire. Initially under British administration, the country became an independent kingdom in 1932. The monarchy was overthrown in a 1958 military coup and a decade later the Arab nationalist Ba'ath Party seized power. Saddam Hussein emerged as the dominant figure and became president in 1979. He led Iraq into an inconclusive war with Iran (1980–88) and waged a genocidal campaign against the Kurdish minority in the north. Ba'ath rule ended in 2003, when an American-led coalition invaded and overthrew Hussein, establishing a democratic constitution. Multi-party elections were held in 2005, but a widespread armed insurgency quickly erupted, necessitating the continued presence of coalition forces. Since their departure in 2011, violence has escalated into an ongoing civil war between government forces and the militant Islamic State.

Republic of Iraq
33.31° N, 44.36° E
UTC + 3

☐	435,050 km² (167,974 sq mi)
○	Mostly desert
♀	37,202,572
⊞	85.7/km² (222/sq mi)
↑↑	3.0%
◻	69.6 : 30.4%
⋔	77% Arab 17% Kurdish 6% others
♟	Arabic, Kurdish
☐	Islam
🏛	Federal parliamentary republic
⊛	AL, G77, IMF, NAM, OPEC, UN, WB, WTO (observer)
🏴	Iraqi Dinar (IQD)
⊞	$171,49 bn
▦	$4,610
☰	Oil

TURKEY

Dahuk

Mosul

Irbil

As Sulaymaniyah

SYRIA

Kirkuk

Tikrit

Djela

IRAN

Ba'Qubah

Euphrates

Tigris R.

★ BAGHDAD
6.6 m

Ar Ramadi

Al Hillah

Al Kut

JORDAN

Karbala

Euphrates

An Najaf

Ad Diwaniyah

Al Amarah

As Samawah

SAUDI ARABIA

An Nasiriyah

Al Basrah

0		100 miles

0		100 km

KUWAIT

PERSIAN GULF

Georgia

From the 11th to the 13th centuries, the Kingdom of Georgia was a major power in the Caucasus region, but it had split apart by late 15th century, and by the 19th century, Russian rule had gradually extended over the region. Following a brief period of independence from 1918–21, it was incorporated into the USSR as the Georgian Soviet Socialist Republic. Seventy years later the country declared independence, becoming a democratic republic in 1991. However, within a few months of the first elections, a coup overthrew the president, provoking a two-year civil war. Separatist forces in the regions of Abkhazia and South Ossetia were drawn into this conflict, with both regions winning de facto independence (though little international recognition). Georgian political leaders retained strong ties with Russia, until the 2003 Rose Revolution brought about a new regime that favoured closer integration with the West. Russo-Georgian relations soon deteriorated; in 2008, a five-day war between the two resulted in Russian military occupation of Abkhazia and South Ossetia.

Georgia
41.72° N, 44.82° E (T'bilisi)
UTC + 4

Georgia has two capitals –
K'ut'aisi (where the legislature
is located) and T'bilisi.

☐ 69,700 km² (26,911 sq mi)	☐ 53.8 : 46.2%	✪ COE, IMF, UN, WB, WTO
○ Warm temperate	♟ 82% Georgian 18% others	⌨ Georgian Lari (GEL)
♱ 3,719,300 (not inc. 54,000 in South Ossetia and 240,000 in Abkhazia)	♟ Georgian, Abkhaz, Ossetian and Russian	▦ $14.33 bn
		▦ $3,854
⊡ 65/km² (168/sq mi)	⌂ Christianity	⚒ Automotive manufacture, mining, agriculture and food processing
♰ 0.1%	⚒ Unitary semi-presidential constitutional republic	

Yemen

Occupying a strategic position at the southern entrance to the Red Sea, Yemen is mostly arid and mountainous. The imposition of central authority has always been problematic in Yemen because of the long-standing importance and power of local tribes. During the 19th century, nominal sovereignty was split between the Ottomans, who claimed the northwest, and the British, who established a protectorate based around the port of Aden, in order to safeguard the trade route to Asia. In 1918, following the dissolution of the Ottoman Empire, its Yemeni territory became independent North Yemen. The area under British control finally gained independence in 1967 as South Yemen; three years later, it became a one-party Marxist-socialist state with close ties to the USSR. Despite occasional hostility, the two states unified to form a single republic in 1990 and remained together following the defeat of southern separatists in a 1994 civil war. Yemen remains the poorest country in the Middle East; in 2011, the country descended into a political crisis, leading to a civil war that began in 2015.

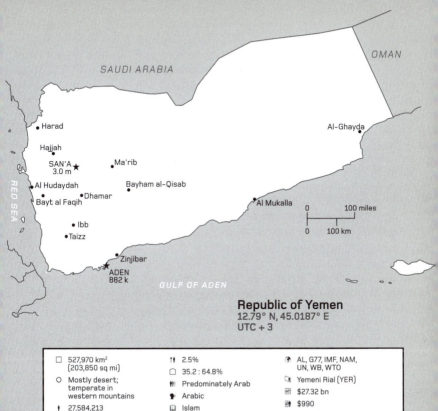

SAUDI ARABIA

OMAN

• Harad

Al-Ghayda •

• Hajjah

SAN'A ★
3.0 m

• Ma'rib

• Al Hudaydah

Bayham al-Qisab •

• Dhamar

• Bayt al Faqih

Al Mukalla •

RED SEA

• Ibb

0		100 miles
0		100 km

• Taizz

• Zinjibar

ADEN
882 k

GULF OF ADEN

Republic of Yemen
12.79° N, 45.0187° E
UTC + 3

☐ 527,970 km² (203,850 sq mi)	♈ 2.5%	◉ AL, G77, IMF, NAM, UN, WB, WTO
O Mostly desert; temperate in western mountains	☐ 35.2 : 64.8%	▧ Yemeni Rial (YER)
♦ 27,584,213	♟ Predominately Arab	▦ $27.32 bn
▣ 52.2/km² (135/sq mi)	♟ Arabic	▦ $990
	☐ Islam	▤ Oil and gas
	♣ Provisional government	

Armenia

First established as a kingdom in 331 BCE, Armenia adopted Christianity as its state religion in the early fourth century; the first country in the world to do so. This historic state ruled a regional empire covering much of Transcaucasia; current Armenian territory is about ten per cent of its size. Subsequently, sovereignty over the area passed between the Byzantines, Persians and Ottomans. Under Ottoman rule, decades of persecution culminated in a 1915 attempted genocide that saw over one million deaths. The country was briefly independent after the Russian Revolution and incorporated into the USSR in 1920. Shortly after independence in 1991, skirmishes with Azerbaijan over its ethnic-Armenian Nagorno-Karabekh region escalated into war. The conflict ended with Armenian victory in 1994. The disputed area won de facto independence, although it continues to be recognized as part of Azerbaijan, internationally. The transition to a free-market democracy has been troubled; in 2008, mass protests were violently repressed following allegations of electoral fraud by the victorious ruling party.

GEORGIA

Kalinino

Uzunlar

Debed

Spitak

Idzhevan

Kumayri
(Leninakan)

Kirovakan

Dilizhan

Sevan

AZERBAIJAN

Charentsavan

Razdan

Ashtarak

Kamo

Echmiadzin

Lake
Sevan

Zod

Oktemberyan

YEREVAN
1 m

Hrazdan

Martuni

Aras

Artashat

TURKEY

Ararat

Ozhermuk

Angekhakot

Vorotan

Goris

AZERBAIJAN

Kafan

Megri

Republic of Armenia
40.18° N, 44.50° E
UTC + 4

☐ 29,740 km² (11,483 sq mi)	🏛 Unitary semi-presidential republic
○ Highland continental	🌐 CIS, COE, EAEU, IMF, UN, WB, WTO
✚ 2,924,816	💵 Armenian Dram (AMD)
🗒 102.7/km² (266/sq mi)	💰 $10.55 bn
↕ 0.3%	📊 $3,606
☐ 62.6 : 37.4%	🏭 Agriculture and food processing, mining (particularly of molybdenum), diamond-processing
👫 98% Armenian 2% others	
👤 Armenian	
📖 Christianity	

0   40 miles

0   40 km

Iran

With a history dating back to the fourth millennium BCE, Iran was the historic location of the Persian Empire, which, at its peak in the sixth century BCE, ruled territory extending from modern-day Pakistan to Eastern Europe. While its size diminished, the empire remained a major power until the early 19th century, when it lost much of its land to Russia after a series of wars.

In 1906, popular pressure led Persia to adopt a democratic system of constitutional monarchy, and in 1935 the country first requested to be called by its native name, Iran. The Iranian monarchy reasserted control through an Anglo-American-backed 1953 coup, but increasingly autocratic rule led to its overthrow in 1979. An Islamic republic followed, with a democratically elected president and parliament, although conservative Shia Muslim clerics hold supreme executive power. The regime weathered demands for reform in the 1990s, and has been the subject of international condemnation for its oppressive nature, plans to develop nuclear weapons and sponsorship of terrorist groups, particularly in Lebanon.

AZERBAIJAN

TURKEY

ARM.

Aras

Tabriz
Ardebil

CASPIAN
SEA

TURKMENISTAN

Zanjan
Rasht

Gorgan

Bojnurd

Mahabad

Qazvin

Sari

Shahrud

Mashhad

IRAQ

Sanandaj

TEHRAN
8.4 m ★

Semnan

Qom

Bakhtarun
Ilam

Arak
Khorramabad

Kashan

Dezful

Esfahan
Shahr-e Kord

Birjand

AFGHANISTAN

Rud-e Karun

Ahvaz

ZAGROS

Yazd

KUWAIT

Yasuj

Kerman

Shiraz

Bandar-e
Bushehr

Sirjan

Bam

Zahedan

SAUDI
ARABIA

BAHRAIN

Kangan

Bandar-e
Abbas

Bazman

PAKISTAN

QATAR

PERSIAN
GULF

Jask

GULF OF
OMAN

ARABIAN
SEA

UAE

OMAN

- 1,745,150 km² (673,806 sq mi)
- Mostly arid or semi-arid
- 80,277,428
- 49.3/km² (128/sq mi)
- 1.1%
- 73.9 : 26.1%
- 61% Persian
  16% Azeri
  10% Kurdish
  13% others
- Persian
- Islam
- Islamic republic
- ACD, G-15, G24, G77, IMF, NAM, OPEC, UN, WB, WTO (observer)
- Iranian Rial (IRR)
- $393.44 bn (2015)
- $4,958 (2015)
- Oil and gas

Islamic Republic of Iran
35.69° N, 51.39° E
UTC + 3.5

0 ———— 240 miles

0 ———— 240 km

Azerbaijan

With the biggest port on the Caspian Sea (Baku) and a strategic position on trade routes from Central Asia to Europe, control of Azerbaijan has always been much coveted. After centuries of Turkic rule, it fell under Persian control from 1501, before becoming part of the Russian empire in 1813. Independent from 1918–20, it was then incorporated into the USSR as the Azerbaijan Soviet Socialist Republic, finally regaining independence in 1991. Within Azerbaijan's borders lies the Republic of Artsakh, formerly Nagorno-Karabakh, with a mostly ethnic-Armenian population. The territory became the subject of conflict between Azerbaijan and Armenia, which the latter won, in 1994. The disputed area gained de facto independence, but is still recognized as part of Azerbaijan. Politics have been coloured by coup attempts, corruption, suspected electoral fraud and an increasingly authoritarian government. However, economic growth has been robust, thanks to revenues from the country's rich reserves of petroleum and natural gas, which have brought a reduction in poverty and improved infrastructure.

Republic of Azerbaijan
40.41° N, 49.87° E
UTC + 4

□ 86,600 km² (33,436 sq mi)	⚔ 92% Azerbaijani 8% others	⊕ CIS, COE, IMF, NAM, UN, WB, WTO (observer)
○ Semi-arid steppe	⚑ Azerbaijani, Armenian (in Nagorno-Karabakh)	⌨ Azerbaijani Manat (AZN)
✝ 9,762,274	▢ Islam	▦ $37.85 bn
▦ 118.1/km² (306/sq mi)	⚙ Unitary dominant-party semi-presidential republic	▦ $3,877
⇈ 1.2%		
◠ 54.9 : 45.1%		⛽ Oil and gas

Map labels:
GEORGIA, RUSSIA, ARMENIA, IRAN, CASPIAN SEA

Belokany, Zakataly, Khudat, Khachmas, Kuba, Akstafa, Kazakh, Sheki, Tauz, Mingechaur, Geokchay, Ismailly, Gyandzha (Kirovabad), Yevlakh, Akhsu, Shemakha, Myusyuslyu, Sumgait, BAKU ★ 2.4 m, Barda, Kura, Kyurdamir, Agdam, Agdzhabedi, Alyat, Stepanakert, Sabirabad, Ali-Bayramly, Lachin, Martuni, Aras, Sal'yany, Nakhichevan, Pushkino, Kafan, Dzhul'fa, Aras, Yardymly, Lerik, Lenkoran

0   50 miles
0   50 km

Kuwait

The emirate of Kuwait is a mostly dry, inhospitable desert. Its capital was founded in the early 17th century, rising from a fishing village and small fort to a major regional centre of boat-building and maritime trade. The Al-Sabah dynasty has ruled since 1756 – initially as an autonomous area within the Ottoman Empire until, in 1899, it became a British protectorate in order to preserve domestic self-rule. Kuwaiti economy and society transformed in 1937, when its vast reserves of crude oil (some six per cent of the global total) were discovered, allowing it to become one of the wealthiest countries in the world.

When the British protectorate ended in 1961, Kuwait became fully independent, enacting a constitution that provided for a democratically elected parliament with the emir retaining executive power. Crisis struck in August 1990 when Iraq invaded and annexed Kuwait, although the invaders were driven out in February 1991 by a UN coalition, and the country's sovereignty was restored.

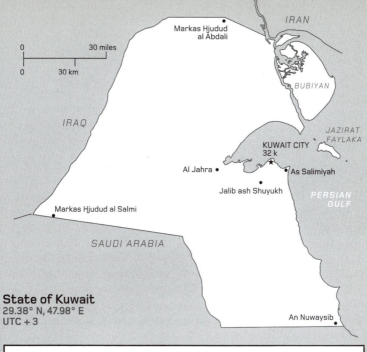

Markas Hjudud
al Abdali

IRAN

BUBIYAN

IRAQ

0 ____ 30 miles
0 ____ 30 km

JAZIRAT
FAYLAKA

KUWAIT CITY
32 k

Al Jahra •

As Salimiyah

Jalib ash Shuyukh

PERSIAN
GULF

Markas Hjudud al Salmi •

SAUDI ARABIA

An Nuwaysib

State of Kuwait
29.38° N, 47.98° E
UTC + 3

☐ 17,820 km² (6,880 sq mi)	⚧ 38% Asian 31% Kuwaiti 28% other Arab 3% others	🌐 ACD, AL, GCC, G77, IMF, NAM, OPEC, UN, WB, WTO
○ Dry desert	☻ Arabic	💱 Kuwaiti Dinar (KWD)
⚲ 4,052,584	▭ Islam	▦ $114.04 bn (2015)
⊡ 227.4/km² (589/sq mi)	⚱ Unitary constitutional monarchy	▨ $28,975 (2015)
⚲⚲ 2.9%		🛢 Oil
◠ 98.4 : 1.6%		

Kazakhstan

The largest landlocked country in the world, Kazakhstan once lay at the centre of the Kazakh Khanate, a union of Muslim nomadic tribes that ruled the modern-day territory as well as much of Uzbekistan. During the late 18th and 19th centuries, the Russian Empire extended its dominion over the area and colonized it. Kazakhstan was briefly independent after the 1917 Russian Revolution, but became part of the USSR in 1920. During the 1950s and 1960s, the Soviet leadership sought to make the area a major grain producer, encouraging migration there, particularly from Russia. As a result, ethnic Kazakhs became a minority. Following the country's independence in 1991, many non-Kazakhs left the country and many ethnic Kazakhs returned, meaning they now make up the majority of the population. The first (and only) president since independence is Nursultan Nazarbayev, who has been the national leader since the communist era. Although the country is officially a democracy, his regime is authoritarian and politically suppressive. The economy is largely based on its extensive reserves of fossil fuels, minerals and metals.

Republic of Kazakhstan
51.16° N, 71.47° E
UTC +5 to UTC +6

☐ 2,724,902 km² (1,052,091 sq mi)	⚲ 63% Kazakh 24% Russian 13% others	☸ ACD, CIS, EAEU, IMF, SCO, UN, WB, WTO
○ Continental	⚲ Kazakh, Russian	⛿ Kazakhstani Tenge (KZT)
⚲ 17,797,032	⚏ Islam	⚏ $133.66 bn
⚏ 6.6/km² (17/sq mi)	⚒ Unitary dominant-party presidential constitutional republic	⚏ $7,510
⚲ 1.4%		⚒ Oil and minerals, metals
☐ 53.2 : 46.8%		

Bahrain

Located in the Persian Gulf, Bahrain is an archipelago of one main island and 30 smaller ones. Bahrain's present reigning dynasty, the Arab Al-Khalifas, conquered the country from the Persian empire in 1783; during the 19th century, it preserved its rule by becoming a British protectorate. The historic foundation of the Bahraini economy was pearl hunting, practised since at least the fifth millennium BCE. This changed in the 1930s, when oil was discovered and the economy shifted towards this industry, leading to rapid modernization. The country declared independence in 1971, with its royal family retaining power through control of executive government, while an elected legislature was also instituted. As Bahrain's oil reserves are limited compared to those of its neighbours, the economy diversified towards other sectors in the 1990s, such as tourism and banking. The country has remained prosperous, although there is still occasional internal strife and protest (such as in 2011 during the Arab Spring uprisings). This stems mainly from the fact that its ruling house is Sunni Muslim, while the majority of the population is Shia.

SAUDI
ARABIA

PERSIAN GULF

Muharraq

MANAMA
411 k

Kingdom of Bahrain
26.2° N, 50.59° E
UTC +3

- ☐ 771 km² (298 sq mi)
- ○ Arid
- ♦ 1,425,171
- ⊞ 1,848.5/km² (4,788/sq mi)
- ♦♦ 3.8%
- ☐ 88.8 : 11.2%
- ♦♦♦ 46% Bahraini Arab
  46% East Asian
  8% others
- ♦ Arabic
- ☐ Islam
- ♨ Unitary presidential system under constitutional monarchy
- ☻ ACD, AL, GCC, G77, IMF, NAM, UN, WB, WTO
- ☒ Bahraini Dinar (BHD)
- ▦ $31.86 bn
- ▦ $22,354
- ☒ Oil

• Riffa

• Awali

GULF OF BAHRAIN

• Mamtalah

SAUDI
ARABIA

0 ——————— 10 miles
0 ——————— 10 km

Qatar

A rectangular peninsula extending into the Persian Gulf from the northeastern coast of Arabia, most of Qatar is flat desert. The indigenous tribes converted to Islam in the early seventh century and a series of different dynasties ruled before the Al-Thani family took control in 1850; they have been emirs ever since, retaining a high degree of autonomy when the country was under the sovereignty of the Ottoman Empire (1871–1915). During the First World War, Qatar joined a regional Arab revolt against Ottoman rule and became a British protectorate, later gaining independence in 1971. The discovery of oil in 1939 signalled a period of rapid economic growth and tremendous prosperity. Income from petroleum and natural gas has allowed the country to become one of the most influential in the Arab world, particularly through the state-owned Al Jazeera media company and numerous foreign investments. Relations with more conservative Arab states, such as Saudi Arabia, are strained, particularly due to Qatar's close ties with Iran and support of reformist movements during the Arab Spring of 2010–12.

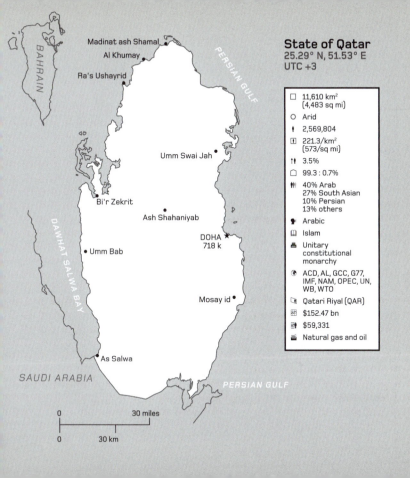

BAHRAIN

Madinat ash Shamal
Al Khumay
Ra's Ushayrid

PERSIAN GULF

Umm Swai Jah

Bi'r Zekrit

Ash Shahaniyab

DOHA
718 k

DAWHAT SALWA BAY

Umm Bab

Mosay id

SAUDI ARABIA

As Salwa

PERSIAN GULF

0 ─── 30 miles

0 ─── 30 km

State of Qatar
25.29° N, 51.53° E
UTC +3

☐ 11,610 km²
  (4,483 sq mi)

○ Arid

🚹 2,569,804

▦ 221.3/km²
  (573/sq mi)

🚹🚹 3.5%

◯ 99.3 : 0.7%

🧑 40% Arab
  27% South Asian
  10% Persian
  13% others

🗣 Arabic

📖 Islam

♟ Unitary
  constitutional
  monarchy

🌐 ACD, AL, GCC, G77,
  IMF, NAM, OPEC, UN,
  WB, WTO

💵 Qatari Riyal (QAR)

🏦 $152.47 bn

🏧 $59,331

⛴ Natural gas and oil

United Arab Emirates

In 1820, seven emirates on the eastern coast of the Arabian Peninsula joined together to sign a treaty with the United Kingdom, creating a polity known as the Trucial States. The area, almost entirely comprising desert, remained a British protectorate until 1971, when it won independence as the UAE. One emirate, Ras al-Khaimah, did not join until the following year, while the nearby states of Bahrain and Qatar declined the chance to join the federation. Each member of the UAE retained its own hereditary dynasty, which ruled domestically as an absolute monarchy. Providing overall leadership is the Federal Supreme Council, comprised of the seven emirs. The president is customarily the emir of Abu Dhabi, the largest and most populous emirate, while the office of prime minister traditionally falls to the emir of Dubai, the second-largest emirate. The UAE's fortunes transformed during the 1950s, when oil was discovered in its territory. Although other economic sectors, particularly transport and tourism, have become significant, oil and natural gas continue to be the main foundation of Emirati wealth and prosperity.

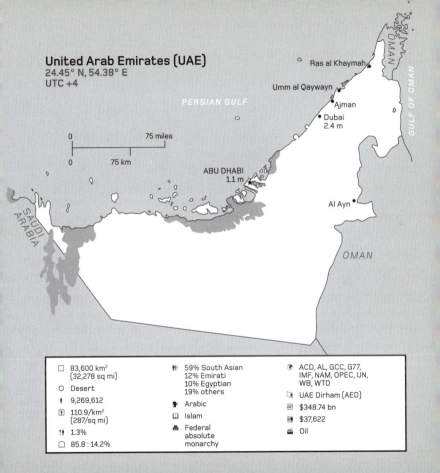

United Arab Emirates (UAE)
24.45° N, 54.38° E
UTC +4

PERSIAN GULF

Ras al Khaymah

Umm al Qaywayn

Ajman

Dubai
2.4 m

ABU DHABI
1.1 m

Al Ayn

OMAN

GULF OF OMAN

SAUDI ARABIA

0   75 miles

0   75 km

☐ 83,600 km² (32,278 sq mi)	59% South Asian 12% Emirati 10% Egyptian 19% others	ACD, AL, GCC, G77, IMF, NAM, OPEC, UN, WB, WTO
○ Desert	Arabic	UAE Dirham (AED)
9,269,612	Islam	$348.74 bn
110.9/km² (287/sq mi)	Federal absolute monarchy	$37,622
1.3%		Oil
85.8 : 14.2%		

Oman

Occupying a strategic position at the mouth of the Persian Gulf, Oman spent 200 years under the dominion of the Ottomans, Portuguese and Persians, successively, before the Arab Al-Said dynasty took control in the mid-18th century. The independent sultanate became an imperial power, ruling the island of Zanzibar and territories on the East African, Iranian and Pakistani coasts. The Omani empire went into decline from the mid-19th century, losing its overseas possessions, although its capital, Muscat, remained one of the most important ports in Asia. Initially, the sultan ruled coastal areas, while religious leaders controlled the autonomous interior. However, by 1970, the sultanate had extended control over the entire country, having defeated rebellions from the interior as well as leftist forces. Since that time, a series of political reforms established an elected parliament, although the sultan retains sweeping powers. As Oman's oil reserves are modest by regional standards, the government has worked to develop a more diversified economy incorporating natural gas, port services and light industry.

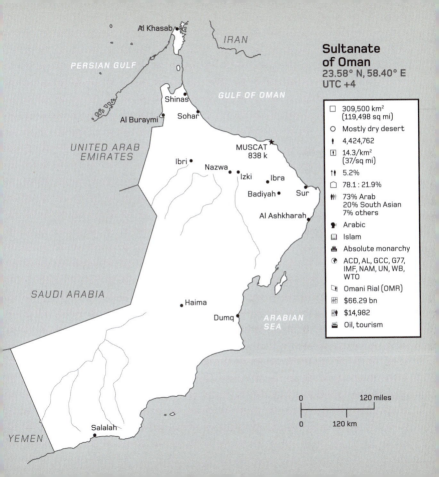

IRAN

PERSIAN GULF

Al Khasab

GULF OF OMAN

Shinas

Al Buraymi

Sohar

UNITED ARAB
EMIRATES

MUSCAT
838 k

Ibri

Nazwa

Izki

Ibra

Badiyah

Sur

Al Ashkharah

SAUDI ARABIA

Haima

Dumq

ARABIAN
SEA

Salalah

YEMEN

Sultanate
of Oman
23.58° N, 58.40° E
UTC +4

☐	309,500 km² (119,498 sq mi)
○	Mostly dry desert
👤	4,424,762
⊞	14.3/km² (37/sq mi)
⚥	5.2%
⌂	78.1 : 21.9%
👪	73% Arab 20% South Asian 7% others
🗣	Arabic
☪	Islam
⚖	Absolute monarchy
◉	ACD, AL, GCC, G77, IMF, NAM, UN, WB, WTO
💱	Omani Rial (OMR)
📊	$66.29 bn
💰	$14,982
🏭	Oil, tourism

0	120 miles
0	120 km

Turkmenistan

The indigenous population of this country are the Turkmen, a Turkic people who converted to Islam after the eighth century. Over 90 per cent of the landscape is sandy desert but, because of its position on the trade route from Asia to Europe, many empires, including the Persians, Arabs and Mongols, have vied for control of the area. Russia extended dominance over the territory during the late 19th century, but faced sporadic armed uprisings that continued after the Russian Revolution. Nevertheless, in 1924, Turkmenistan became part of the USSR as the Turkmen Soviet Socialist Republic. Soviet rule was authoritarian, with local culture repressed and the region kept highly isolated. The economy shifted from mainly nomadic pastoral agriculture to focusing on growing raw cotton, as well as tapping the country's rich deposits of natural gas. When the USSR dissolved in 1991, Turkmenistan became independent, with the leader of the local communist party, Saparmurat Niyazov, ruling until his death in 2006. The country remains highly politically repressive, and any organized opposition to the ruling party is suppressed.

Turkmenistan
37.96° N, 58.33° E
UTC + 5

☐ 488,100 km² (188,456 sq mi)	♛ 85% Turkmen 15% others	✪ CIS (associate), G77, IMF, NAM, UN, WB
○ Subtropical desert	♟ Turkmen	⌨ Turkmen New Manat (TMT)
⭑ 5,662,544	⌂ Islam	▦ $36.18 bn
⊞ 12/km² (31/sq mi)	⚏ Unitary dominant-party presidential republic	⛁ $6,389
⇅ 1.7%		⛴ Natural gas and oil, cotton
⌂ 50.4 : 49.6%		

Uzbekistan

Historically a central part of the Transoxiana region of Central Asia, Uzbekistan has experienced rule under the Persians, Greeks, Chinese, Arabs, Mongols and Turks. The Russians conquered Uzbek territory during the late 19th century and, despite local resistance against communist rule, it became part of the USSR as the Uzbek Soviet Socialist Republic in 1924.

Soviet leadership increasingly shifted the main economic focus to cotton, but a reckless drive to maximize production caused serious long-term environmental problems, with heavy use of agrochemicals leaving the soil polluted. The overuse of fresh water for irrigation has dried up rivers, while the Aral Sea has shrunk to half its original size since the 1960s. Following independence in 1991, the government has diversified by taking advantage of oil, natural gas and mineral reserves. Although the country is officially a democracy, elections are not free; the ruling regime is authoritarian and regularly violates human rights – for example, through forced labour in the cotton industry.

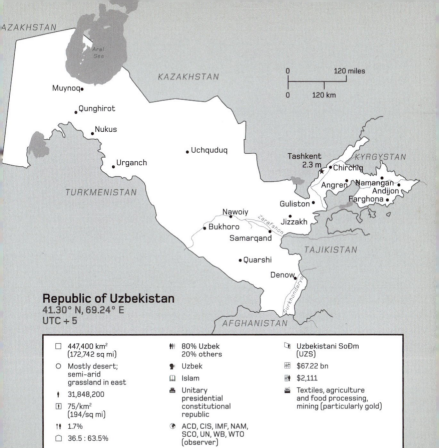

Republic of Uzbekistan
41.30° N, 69.24° E
UTC + 5

☐	447,400 km² (172,742 sq mi)	♂	80% Uzbek 20% others	🗪	Uzbekistani SoĐm (UZS)
○	Mostly desert; semi-arid grassland in east	♟	Uzbek	🎟	$67.22 bn
✝	31,848,200	📖	Islam	📊	$2,111
⊞	75/km² (194/sq mi)	♨	Unitary presidential constitutional republic	🚂	Textiles, agriculture and food processing, mining (particularly gold)
⇈	1.7%	⊛	ACD, CIS, IMF, NAM, SCO, UN, WB, WTO (observer)		
☐	36.5 : 63.5%				

Afghanistan

With an imposing landscape of rugged mountains and desert, Afghanistan is a multi-ethnic country in which tribal affiliation remains a powerful force. The modern state coalesced in the 18th and 19th centuries, when the Pashtun Hotak and Durrani dynasties established sovereignty over the area. Resisting colonial dominance, it became an independent kingdom in 1926. The monarchy was overthrown in 1973 and a republic proclaimed; five years later a communist coup installed a Soviet-backed dynasty. Opposition from the Mujahideen, anticommunist Islamist guerrillas, provoked a Soviet invasion in 1979, but after their withdrawal ten years later the communist government collapsed. Civil war ensued, during which the Islamic fundamentalist Taliban was founded by hard-line religious students; two years later it captured Kabul and seized power. In the aftermath of the 9/11 attacks on the United States, the Taliban was accused of sheltering al-Qaeda leadership, resulting in a NATO-led invasion to topple the regime. A new interim government was installed in June 2002 and democratic elections have been held since 2004, but Taliban insurgents remain active .

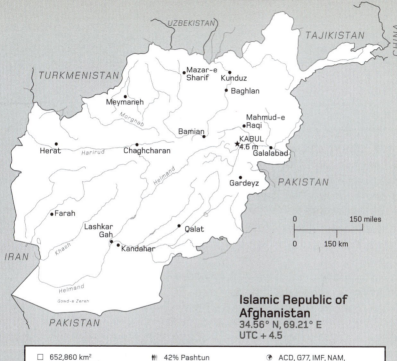

Islamic Republic of Afghanistan
34.56° N, 69.21° E
UTC + 4.5

□ 652,860 km² (252,071 sq mi)	⚑ 42% Pashtun 27% Tajik 31% others	◉ ACD, G77, IMF, NAM, SAARC, UN, WB, WTO
○ Arid to semi-arid	⚑ Pashto, Dari	⌨ Afghani (AFN)
⚑ 34,656,032	☐ Islam	⊞ $19.47 bn
⊞ 53.1/km² (138/sq mi)	♣ Unitary presidential Islamic republic	⊞ $562
⚑⚑ 2.7%		⚎ Agriculture
☐ 27.1 : 72.9%		

Labels on map: UZBEKISTAN, TAJIKISTAN, CHINA, TURKMENISTAN, Mazar-e Sharif, Kunduz, Baghlan, Meymaneh, Morghab, Mahmud-e Raqi, Bamian, KABUL 4.6 m, Herat, Harirud, Chaghcharan, Helmand, Galalabad, Gardeyz, PAKISTAN, Farah, Lashkar Gah, Qalat, Kandahar, Khash, IRAN, Helmand, Gowd-e Zereh, PAKISTAN

Pakistan

The Indus River, once the location of an ancient civilization that dates back to 3300 BCE, flows through the centre of Pakistan. Over the centuries, the area has been invaded and settled by a succession of other peoples, including the Indo-Aryans, Persians, Greeks, Arabs, Turks and Afghans, before coming under British dominion in the 18th century. When the Indian subcontinent gained independence in 1947, it was separated into two states, with Pakistan (which means 'land of the pure') created in Muslim-majority areas. The process of partition was deeply traumatic, with millions displaced and thousands killed. It created lasting enmity with India over the disputed region of Kashmir, which has led to warfare in 1947-48, 1965, 1971 and 1999. Initially, Pakistan was comprised of two sections separated by over 1,600 km (1,000 miles); in 1971, the eastern region became independent as Bangladesh. Although established as a parliamentary democracy, Pakistan's first national elections were not held until 1970, and the country has endured repeated periods of military rule, as well as struggling with the influence of Islamic fundamentalism.

Islamic Republic of Pakistan
33.73° N, 73.09° E
UTC +5

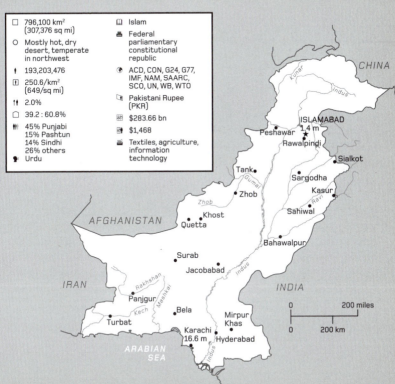

☐ 796,100 km² (307,376 sq mi)	☖ Islam
○ Mostly hot, dry desert, temperate in northwest	♜ Federal parliamentary constitutional republic
♱ 193,203,476	◈ ACD, CON, G24, G77, IMF, NAM, SAARC, SCO, UN, WB, WTO
▣ 250.6/km² (649/sq mi)	
♊ 2.0%	▧ Pakistani Rupee (PKR)
◻ 39.2 : 60.8%	▦ $283.66 bn
♙ 45% Punjabi 15% Pashtun 14% Sindhi 26% others	▤ $1,468
♟ Urdu	▥ Textiles, agriculture, information technology

CHINA

Kunar
Indus

ISLAMABAD
1.4 m
★
Peshawar •
Rawalpindi •
Sialkot •

Tank •
Gumal
Sargodha •
Kasur •
Zhob •
Ravi

Zhob
Sahiwal •

AFGHANISTAN
Khost •
Quetta •

Bahawalpur •

Surab •
Jacobabad •
Indus

IRAN
Rakhshan
Mashkel
Panjgur •
Kech
Bela •
Mirpur Khas •
INDIA

Turbat •
Karachi
16.6 m
Hyderabad •

ARABIAN SEA
Indus

0	200 miles
0	200 km

Tajikistan

The mountainous country of Tajikistan spent much of its history dominated by foreign imperial powers including the Persians, Arabs and Mongols. The majority of the population are Tajik, an ethnic group of Iranian origin who speak a form of Persian. Russia conquered the area in the late 19th century and faced much resistance, which continued after the Russian Revolution. However, by 1924, Tajikistan was incorporated into the Soviet Union, first as part of Uzbekistan but, from 1929, as the separate Tajik Soviet Socialist Republic. The country became independent in 1991, but rapidly descended into a five-year civil war between the government and an opposition coalition of Islamists and democratic reformers. During the conflict the sparsely populated Gorno-Badakhshan region, which makes up around half of the country's area, declared independence; ultimately, however, it remained part of the country as an autonomous area. A ceasefire was concluded in 1997, but sporadic outbreaks of violence continue. The government has since entrenched its power, becoming increasingly authoritarian and marginalizing opposition.

Republic of Tajikistan
38.56° N, 68.79° E
UTC + 5

☐	141,376 km² (54,586 sq mi)	👭	84% Tajik 14% Uzbek 2% others	⌖	ACD, CIS, G77, IMF, SCO, UN, WB, WTO
O	Mid-latitude continental	👤	Tajik	📑	Tajikistani Somoni (TJS)
✝	8,734,951	📖	Islam	💰	$6.95 bn
⊡	62.9/km² (163/sq mi)	⚒	Unitary dominant-party presidential republic	💵	$796
✝✝	2.2%			⛏	Mining, metals, agriculture
⌂	26.9 : 73.1%				

Map labels: UZBEKISTAN, Obanbori Qayroqqum, Khujand, Konibodom, Isfara, UZBEKISTAN, Ayni, KYRGYZSTAN, 75 miles, 75 km, Panjakent, Surkhob, Kyzyl-Suu, Qarokul, CHINA, Garm, DUSHANBE 822 k, Orjonikidzeobod, Tursunzoda, Pani, Norak, Bartang, Murghob, Kulob, Oqsu, Qurghonteppa, Khorugh, Nizhniy Pyandzh, AFGHANISTAN, AFGHANISTAN

India

One of the world's most culturally diverse countries, India has a landscape ranging from rugged mountains and fertile plains to arid desert. Its history has been shaped by periodic invasions from the north, starting with the Indo-Aryan tribes from 1500 BCE, and including Arabs, Turks and Persians. Occasionally, imperial powers would rise, but none imposed dominance over all of India's present territory. The British, however, extended control over the entire Indian subcontinent from the mid-19th century, and continued to rule until 1947, against decades of nationalist campaigning and nonviolent resistance. At this time, Muslim-majority areas were partitioned to form the separate state of Pakistan. Today, the Indian nation is the world's largest democracy (and second most populous nation), with a federal system of 29 states, each with a high degree of autonomy and its own elected government, and seven 'union territories' with slightly less independence. Although poverty and corruption are major problems, reforms since 1991 have led to a rapid increase in prosperity and development levels.

Republic of India
28.61° N, 77.21° E
UTC +5.5

Srinagar
Jammu
Amritsar
Simla
PAKISTAN
H I M A L A Y A S
CHINA
Tinsukia
NEW DELHI · Delhi
258 k
NEPAL
BHUTAN
Kohima
Jaisalmer
Agra
Lucknow
Gorakhpur
Dispur
Imphal
Jaipur
Gwalior
Patna
Jodhpur
Kota
Allahabad
Ganges
Benares
BANGLADESH
Agartala
MYANMAR
Kandla
Ahmadabad
V I N D H Y A S
Bhopal
Jabalpur
Calcutta
Aizawl
Indore
Narmada
Jamshedpur
Surat
Nagpur
Raipur
Mahanadi
Balasore
Mumbai
12.0 m
Godavari
Bhubaneswar
Pune
Vishakhapatnam
BAY
OF
BENGAL
Krishna
Hyderabad
Panaji
W E S T E R N
Guntakal
Chennai
(Madras)
ANDAMAN
ISLANDS
Bangalore
Port Blair
G H A T S
Pondicherry
Calicut
ANDAMAN
SEA
Cochin
Madurai
INDIAN OCEAN
Trivandrum
SRI
LANKA
NICOBAR
ISLANDS

Jamuna
Ganges

☐	3,287,259 km² (1,269,218 sq mi)
○	Varied; tropical monsoon in south, temperate in north
𝕚	1,324,171,354
⊡	445.4/km² (1,156/sq mi)
↟↟	1.1%
⋔	33.1 : 66.9%
👥	72% Indo-Aryan 25% Dravidian 3% others
🗣	Hindi, English, as well as 21 other 'scheduled languages' that have special status
📖	Hinduism
🏛	Federal parliamentary constitutional socialist republic
⊕	ACD, CON, G-15, G24, G77, IMF, NAM, SAARC, SCO, UN, WB, WTO
💱	Indian Rupee (INR)
📊	$2.26 tn
📈	$1,709
🏭	Textiles, agriculture and food processing, information technology

0 _____ 400 miles
0 _____ 400 km

Kyrgyzstan

The Kyrgyz are a Muslim Turkic people who came under Mongol dominance in the 13th century, and whose territory was later conquered by the Chinese empire and the Uzbeks. The Russian empire annexed most of Kyrgyzstan in 1876, but faced repeated revolts. The most widespread occurred in 1916, when around one-sixth of the population was killed. In 1919, the area was added to the USSR as an autonomous province; in 1936, it became the Kirghiz Soviet Socialist Republic. The country declared independence in 1991, with an important local communist official, Askar Akayev, becoming its first president. During the 2005 Tulip Revolution – a mass protest inspired by poverty, unemployment and allegations of electoral fraud – Akayev resigned and fled the country. Although Kyrgyzstan has seen considerable political turmoil and some violence since then, the country's 2011 elections saw the first peaceful transition of power in Kyrgyz history. Despite this political progress, and owing to its historic dependence on trade with the USSR, Kyrgyz economic growth has been slow and the country remains highly reliant on agriculture.

Kyrgyz Republic
42.88° N, 74.57° E
UTC + 6

KAZAKSTAN

Kirov

Talas

BISHKEK
865 k

Tokmok

Chelpori-Ata

Karakol
(Przheval'sk)

Balykchy (Issyk-Kul)

Ysyk-Köl

Chatkal

Kara-Kol

Chaek

Tash-Komur

Naryn

Kara-Say

UZBEKISTAN

Jalal-
Abad

Naryn

Syrdaryo

At-Bashy

Osh

Ak-Say

Kyzyl-Kyya

Suluktu

CHINA

Khaydarkan

Sary-Tash

150 miles

150 km

TAJIKISTAN

▢ 199,949 km² (77,201 sq mi)	⥮ 2.1%	◉ ACD, CIS, EAEU, IMF, SCO, UN, WB, WTO
○ Dry continental to polar in high Tien Shan Mountains, subtropical in southwest temperate in northern foothills	⌂ 35.9 : 64.1%	🏷 Kyrgyzstani Som (KGS)
	⚥ 71% Krygyz 14% Uzbek 15% other	💰 $6.55 bn
	👤 Kyrgyz, Russian	💵 $1,077
● 6,082,700	📖 Islam	⛏ Agriculture, mineral mining (particularly gold)
⊞ 31.7/km² (82/sq mi)	🏛 Unitary parliamentary republic	

Maldives

A chain of 26 atolls in the Indian Ocean, the Maldives comprise 1,200 islands and sandbanks stretching over 800 km (500 miles) north to south. The area was first settled in around 1500 BCE by people from southern India and Sri Lanka, who were joined by Indo-Aryans about 1,000 years later. By the 12th century, the islands had come under the rule of a sultan based on the central island of Malé; at this time the Maldivian people adopted Islam, introduced by Asian and Persian traders.

Maldives was a British protectorate from 1887 until 1965, when it regained independence with the sultan continuing as a constitutional monarch. Following a referendum in 1968, however, the sultanate was abolished and a new republican system established. The country remains one of the least industrialized in Asia, with much of the population involved in fishing or farming, although tourism has become increasingly important. The Maldives' most serious challenge is environmental; as the world's lowest-lying nation it is highly vulnerable to rising sea levels.

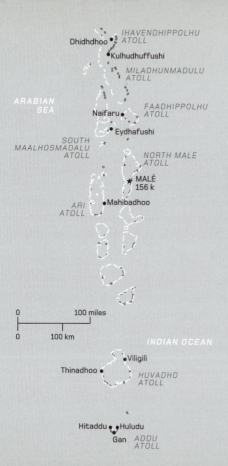

IHAVENDHIPPOLHU
ATOLL
Dhidhdhoo

Kulhudhuffushi

MILADHUNMADULU
ATOLL

ARABIAN
SEA

FAADHIPPOLHU
ATOLL
Naifaru

Eydhafushi

SOUTH
MAALHOSMADALU
ATOLL

NORTH MALE
ATOLL

★ MALÉ
156 k

ARI
ATOLL

Mahibadhoo

0   100 miles

0   100 km

INDIAN OCEAN

Viligili

Thinadhoo

HUVADHO
ATOLL

Hitaddu Huludu

Gan   ADDU
ATOLL

Republic
of Maldives
4.18° N, 73.51° E
UTC + 5

☐	300 km² (116 sq mi)
○	Tropical
✝	417,492
⊞	1,391.6/km² (3,694/sq mi)
↟↟	2.0%
◠	46.5 : 53.5%
⚲	94% Maldivian 6% others
☻	Maldivian
📖	Islam
🏛	Unitary presidential constitutional republic
◉	G77, IMF, NAM, SAARC, UN, WB, WTO
🗋	Maldivian Rufiyaa (MVR)
💹	$3.59 bn
💴	$8,602
🚃	Tourism, fishing

China

The Chinese empire was founded in 221 BCE from the unification of seven former 'Warring States'. Under imperial rule for over two millennia, the realm steadily grew in size and became probably the wealthiest and most advanced in the world. This changed in the 19th century, when internal conflict, famine and foreign incursions destabilized the country, leading to the overthrow of the emperor and the establishment of a republic in 1912. Civil war between communists and nationalists began in 1927 and fighting continued until 1950, with the communists eventually triumphing, and declaring the People's Republic of China under the leadership of Communist Party Chairman Mao Zedong in 1949.

The early years of communist rule were turbulent. However, from 1978, reforms began to open up the country, pulling millions out of poverty. With the largest military and highest population in the world, as well as its burgeoning economic might, China is fast becoming a 21st-century superpower.

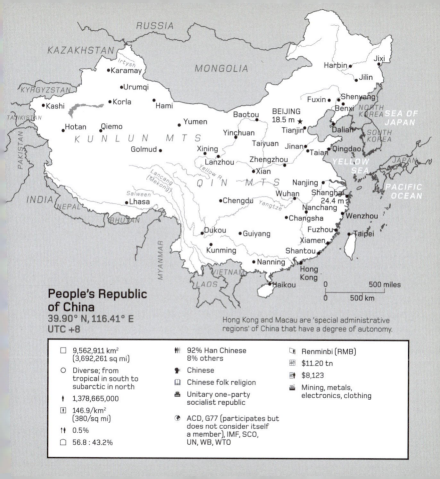

RUSSIA

KAZAKHSTAN

KYRGYZSTAN

MONGOLIA

•Karamay

Irtysh

•Urumqi

TAJIKISTAN

•Kashi

•Korla

•Hami

Harbin•

Jixi•

Jilin•

Fuxin•

Shenyang•

Benxi•

NORTH KOREA

•Hotan

•Qiemo

K U N L U N M T S

•Yumen

Baotou•

BEIJING
18.5 m ★

Tianjin•

Dalian•

SEA OF JAPAN

•Golmud

Yinchuan•

Xining•

Taiyuan•

Jinan•

Taian•

Qingdao•

SOUTH KOREA

JAPAN

Lanzhou•

Zhengzhou•

YELLOW SEA

PAKISTAN

INDIA

NEPAL

Lhasa•

Lancang
(Mekong)

Salween

Yellow R

Q I N M T S

Xian•

•Chengdu

Yangtze

Nanjing•

Wuhan•

Shanghai
24.4 m

PACIFIC OCEAN

BHUTAN

Nanchang•

•Changsha

Wenzhou•

MYANMAR

•Dukou

•Guiyang

Fuzhou•

Xiamen•

Taipei•

Kunming•

Shantou•

LAOS

VIETNAM

•Nanning

Hong Kong

Haikou•

0   500 miles

0   500 km

People's Republic of China
39.90° N, 116.41° E
UTC +8

Hong Kong and Macau are 'special administrative
regions' of China that have a degree of autonomy.

☐ 9,562,911 km² (3,692,261 sq mi)	♀ 92% Han Chinese 8% others	Renminbi (RMB)
○ Diverse; from tropical in south to subarctic in north	☻ Chinese	$11.20 tn
⚤ 1,378,665,000	☐ Chinese folk religion	$8,123
⊞ 146.9/km² (380/sq mi)	⚖ Unitary one-party socialist republic	Mining, metals, electronics, clothing
⚤ 0.5%		
☐ 56.8 : 43.2%	☝ ACD, G77 (participates but does not consider itself a member), IMF, SCO, UN, WB, WTO	

Sri Lanka

The Buddhist Sinhalese people originated in India and migrated to Sri Lanka in the sixth century BCE. Hindu Tamil people from southern India started joining them in the ninth century CE, but no single ruler dominated the shifting array of kingdoms spread across the island. Drawn by the spice trade, Portugal ruled coastal areas in the first half of the 17th century, before being supplanted by the Dutch and then the British. In 1815, the entire island became a British colony called Ceylon (an English transliteration of the Portuguese name, Ceilão). Independence was won in 1948 and, in 1972, the country became a republic, officially changing its name to Sri Lanka. Tensions between the two main ethnic groups led to civil war between the Sinhalese-majority government and Tamil nationalist separatists (1983–2009). The government prevailed and has set about an extensive reconstruction of the country, resettling displaced peoples and diversifying the economy away from being based on agricultural goods like tea, rubber and coffee towards other sectors such as clothes manufacturing, tourism and information technology.

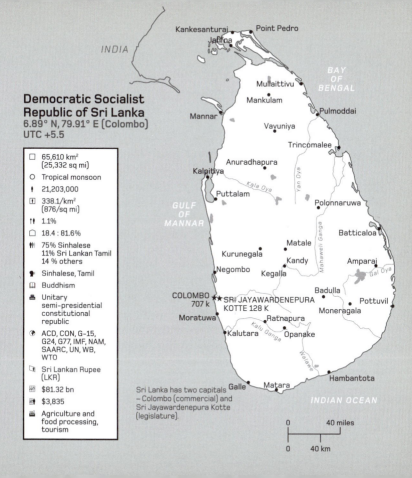

Democratic Socialist Republic of Sri Lanka
6.89° N, 79.91° E (Colombo)
UTC +5.5

- ☐ 65,610 km² (25,332 sq mi)
- ○ Tropical monsoon
- ♀ 21,203,000
- ⊞ 338.1/km² (876/sq mi)
- ♀♀ 1.1%
- ◠ 18.4 : 81.6%
- ♯ 75% Sinhalese 11% Sri Lankan Tamil 14 % others
- ♟ Sinhalese, Tamil
- ▦ Buddhism
- ▦ Unitary semi-presidential constitutional republic
- ◈ ACD, CON, G-15, G24, G77, IMF, NAM, SAARC, UN, WB, WTO
- ▧ Sri Lankan Rupee (LKR)
- ▦ $81.32 bn
- ▦ $3,835
- ▦ Agriculture and food processing, tourism

INDIA

Kankesanturai
Jaffna
Point Pedro

Mullaittivu
Mankulam
Pulmoddai

Mannar
Vavuniya
Trincomalee

BAY OF BENGAL

Anuradhapura
Kalpitiya
Kala Oya
Yan Oya

Puttalam
Polonnaruwa

GULF OF MANNAR

Batticaloa

Kurunegala
Matale
Mahaweli Ganga

Negombo
Kandy
Amparai
Gal Oya

Kegalla
Badulla

COLOMBO
707 k
SRI JAYAWARDENEPURA KOTTE 128 K
Pottuvil

Moratuwa
Ratnapura
Moneragala

Kalutara
Opanake
Kalu Ganga

Galle
Walawe
Hambantota

Matara

INDIAN OCEAN

Sri Lanka has two capitals – Colombo (commercial) and Sri Jayawardenepura Kotte (legislature).

0 ——— 40 miles
0 ——— 40 km

Nepal

The Himalayan country of Nepal contains eight of the ten highest mountains in the world. Its present territory was unified by the Shah dynasty in 1768, which extended its rule into northern India. Nepalese independence was retained even after defeat in the Anglo-Nepalese War (1814–16). However, it lost its Indian territories and, as part of the peace treaty, the British were permitted to recruit soldiers in Nepal (which they continue to do). Democratic rule, established in 1951 with the encouragement and support of India, was short-lived as the king outlawed political parties in 1960. In the face of popular pressure, democracy was restored in 1990 and a constitutional monarchy established. In 1996, a civil war erupted between the government and communist insurgents, who wanted to abolish the monarchy entirely. The fighting ended in 2006; two years later Nepal was declared a federal republic, abolishing the monarchy. Peace has led to some measure of stability, although the country remains one of the poorest in Asia and heavily reliant on remittances from abroad.

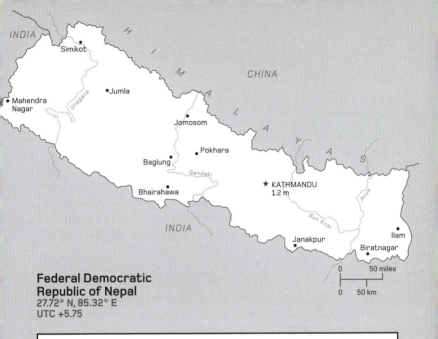

INDIA

HIMALAYAS

•Simikot

CHINA

Ghagara

•Jumla

•Mahendra Nagar

Jomosom

•Pokhara

Baglung

Gandaki

★ KATHMANDU
1.2 m

Bhairahawa

Arun

INDIA

Sun Kosi

Janakpur

Ilam

Biratnagar

0   50 miles

0   50 km

Federal Democratic Republic of Nepal
27.72° N, 85.32° E
UTC +5.75

☐ 147,180 km² (56,827 sq mi)	⌂ 19.0 : 81.0%	⊕ ACD, G77, IMF, NAM, SAARC, UN, WB, WTO
○ Mixture of tropical, temperate, cold, subarctic and arctic	♛ 17% Chhettri 12% Brahman-Hill 71% others	▭ Nepalese Rupee (NPR)
♀ 28,982,771	♙ Nepali	▦ $21.14 bn
⊡ 202.2/km² (524/sq mi)	⌺ Hinduism	▧ $730
♂♀ 1.1%	♜ Federal parliamentary republic	⚒ Agriculture, tourism

Mongolia

The Mongols are a nomadic people who rose to prominence in the 13th century, following a series of conquests across Eurasia that established the largest contiguous empire in world history. After the empire collapsed and fragmented in the 14th century, many Mongols returned to their homeland, which was wholly incorporated into imperial China by 1691. Independence was declared in 1911, but was not fully secured until a decade later. However, the majority of the ethnic Mongol population continues to remain within the Chinese-ruled Inner Mongolia Autonomous Region. In 1924, the communist Mongolian People's Republic was established as a Soviet satellite state; it was overthrown in a peaceful revolution in 1990. Multi-party democratic rule followed, and economic reforms introduced a free-market system. Traditional livestock herding continues, but other sectors, such as mineral exports, have become important, and the economy still relies heavily on China and Russia. Flat steppe and desert remain the most common features of the Mongolian landscape, and the country has the lowest population density in the world.

□ 1,564,170 km²
(603,910 sq mi)

○ Desert

† 3,027,398

⊡ 1.9/km²
(5/sq mi)

†† 1.7%

◻ 72.8 : 27.2%

†† 82% Khalkh Mongols
18% others

♞ Mongolian

📖 Buddhism

♨ Unitary
semi-presidential
republic

✈ ACD, G77, IMF, NAM,
UN, WB, WTO

💵 Mongolian Tögrög
(MNT)

💹 \$11.16 bn

💹 \$3,686

⛏ Mining (particularly
copper), textiles

Mongolia
47.89° N, 106.91° E
UTC +7 to UTC +8

Bangladesh

Until the mid-18th century, the Bengal region of the eastern Indian subcontinent, largely located on the Bengal Delta, was an international trading power and the world's leading producer of textiles. Bengali power declined under British rule that began in 1757 and lasted for nearly two centuries; the region rapidly deindustrialized and suffered several major famines. Under the 1947 Partition Plan, West Bengal joined India, while Muslim-majority East Bengal became part of Pakistan, despite its geographical separation. Demands for greater autonomy and cultural recognition led to a 1971 war of liberation, after which East Pakistan became independent Bangladesh. The country struggled with famine and natural disasters, as well as frequent military coups. Politically, it has stabilized since 1991, with democratic elections held regularly. Despite problems of corruption and weak infrastructure, the Bangladeshi economy has made great strides forward, developing a burgeoning export-led industrial sector, largely based on garment manufacture. The country has seen a steady reduction in poverty in recent decades.

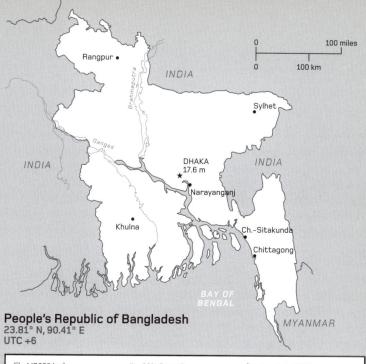

People's Republic of Bangladesh
23.81° N, 90.41° E
UTC +6

☐	147,630 km² (57,000 sq mi)	👥	98% Bengali 2% others	🗋	Bangladeshi Taka (BDT)
O	Tropical	👤	Bengali	💹	$221.42 bn
⬛	162,951,560	📖	Islam	💷	$1,359
⬛	1,251.8/km² (3,242/sq mi)	⚒	Unitary parliamentary republic	🏭	Clothing manufacture, agriculture
👥	1.1%	🌐	ACD, CON, G77, IMF, NAM, SAARC, UN, WB, WTO		
☐	35 : 65%				

Bhutan

Situated on the southern slopes of the Himalayas between Tibet and India, Bhutan was unified by a Buddhist spiritual leader in the 1630s. The country fragmented during the 18th and 19th centuries, then reunited under the leadership of a local governor who became hereditary monarch in 1907. In 1910, the British empire was granted control over Bhutanese foreign affairs in return for respecting its domestic independence; in 1949, Bhutan signed a similar treaty with India that remained in place until 2007. During the first half of the 20th century, Bhutan was an isolated absolute monarchy, but a series of reforms starting in the 1950s began to democratize the country, culminating in the establishment of a constitutional monarchy in 2008. The first national elections were completed that year, and the transition to multi-party democracy has been successful. Bhutanese leaders have been careful to preserve the country's environment and culture; television and the Internet were not allowed until 1999. There is also a national commitment to remaining carbon neutral while developing the economy using hydroelectricity.

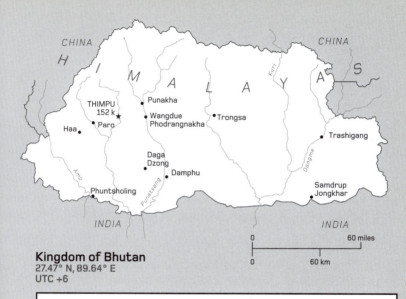

Kingdom of Bhutan
27.47° N, 89.64° E
UTC +6

☐ 38,394 km² (14,824 sq mi)	†† 1.3%	✈ ACD, G77, IMF, NAM, SAARC, UN, WB, WTO (observer)
○ Tropical in southern plains; cool winters and hot summers in central valley; severe winters and cool summers in Himalayas	☐ 39.4 : 60.6%	💵 Bhutanese Ngultrum (BTN), fixed to the value of the Indian Rupee
	††† 50% Ngalop 35% Nepalese 15% indigenous	
† 797,765	🗣 Dzongkha	💲 $2.24 bn
⊞ 20.9/km² (54/sq mi)	☐ Buddhism	🖩 2,804
	♣ Unitary parliamentary system under constitutional monarchy	🌾 Agriculture, forestry

Myanmar

In the 11th century, most of Myanmar's present territory was unified under the leadership of monarchs from the Bamar people, the country's dominant ethnic group. Although it occasionally fragmented, the area, then known as Burma, remained independent until annexation by the British after a series of three wars from 1824–86. Independence as a democracy was achieved in 1948, but an army-led coup overthrew the elected leader in 1962, instituting a brutal and autocratic military dictatorship. The regime changed the country's name to Myanmar in 1989, and in 2005 moved the capital from its historic location of Yangon to Naypyidaw, a planned city in a remote interior location. Although the opposition won elections in 1990, the results were ignored and the military junta continued until 2011, when a democratic, civilian-led government took power. Due to its multi-ethnic population, Myanmar has experienced numerous local armed insurgencies since independence. Tragically, discrimination against minorities continues under the democratic regime, with the Rohingya Muslims in the west of the country facing brutal persecution.

Republic of the Union of Myanmar
19.76° N, 96.08° E
UTC +6.5

☐	676,590 km² (261,233 sq mi)
○	Tropical monsoon
✝	52,885,223
⊡	81/km² (210/sq mi)
↿↾	0.9%
⌂	34.6 : 65.4%
⚤	68% Bamar 32% others
☌	Burmese
☐	Buddhism
⚖	Unitary parliamentary constitutional republic
⊕	ACD, ASEAN, G77, IMF, NAM, UN, WB, WTO
☐	Burmese Kyat (MMK)
☐	$67.43 bn
☐	$1,275
☐	Agriculture, forestry

CHINA

INDIA

Chindwin

Myitkyina

Irrawaddy

BANGLADESH

Maymyo

Mandalay

Pakokku

Taunggyi

Mekong

LAOS

Sittwe

Minbu

NAYPYIDAW
1 m

Salween

Pye

Toungoo

Sandoway

BAY OF
BENGAL

Pegu

Bassein

Moulmein

Amherst

Yangon
4.8 m

THAILAND

Tavoy

ANDAMAN
SEA

Mergui

0	120 miles
0	120 km

Indonesia

An environmentally diverse archipelago spanning one-eighth of Earth's circumference, Indonesia is made up of more than 17,000 islands, within which Java is home to around half the population. Commerce between the islands, as well as with foreigners (the Chinese, Indians and Arabs being particularly influential) has been a central feature of Indonesian history. The Netherlands, attracted by the spice trade, extended colonial dominance over the islands during the 17th and 18th centuries, establishing the Dutch East Indies in 1800. The Japanese occupation of 1942–45 ended Dutch rule and, in the closing weeks of the Second World War, Indonesian nationalist leaders declared the country's independence. Despite attempting to re-establish their rule, the Dutch were forced to recognize Indonesian independence four years later. Authoritarian leadership from 1957–98 gave way to a more liberal, democratic regime. Today, the Indonesian economy is the largest in Southeast Asia, thanks to its plentiful natural resources of minerals, oil and natural gas, as well as its position on international shipping lanes.

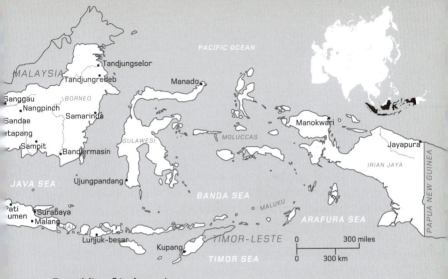

Republic of Indonesia
6.18° S, 106.87° E
UTC +7 to UTC +9

☐	1,910,931 km² (737,815 sq mi)	♈	40% Javanese 16% Sundanese 44% others	✪	ACD, ASEAN, G-15, G77, IMF, NAM, UN, WB, WTO
○	Tropical	☻	Indonesian	▣	Indonesian Rupiah (IDR)
✦	261,115,456	▯	Islam	▦	$932.26 bn
▦	144.1/km² (373/sq mi)	♨	Unitary presidential constitutional republic	▥	$3,570
↟	1.1%			☶	Oil and natural gas, textiles
◠	54.5 : 45.5%				

Thailand

Extending from the northern part of the Malay Peninsula to the centre of mainland Southeast Asia, Thailand's landscape ranges from sandy coastal areas to hilly forests. The Kingdom of Ayutthaya (Siam) began its rise to dominance over Thailand and surrounding areas in 1351, but collapsed in 1767 after a Burmese invasion. The current royal dynasty, the Chakri, came to power in 1782. Despite being sandwiched between French and British colonies, the country remained independent. A bloodless revolution in 1932 saw absolutist royal rule replaced with a system of constitutional monarchy, and subsequently the country has alternated between periods of military rule and democracy, though always matched with fierce loyalty to the crown. In southern provinces the country suffers from ongoing armed insurgency by Muslim Malay separatists. Despite this political instability, in recent decades Thailand has experienced sustained growth with an increasingly industrialized economy, a flourishing tourist industry, sound infrastructure and growing levels of prosperity and development.

Kingdom of Thailand

13.76° N, 100.50° E
UTC +7

- ☐ 513,120 km²
  (198,117 sq mi)
- ○ Tropical
- ♗ 68,863,514
- ▦ 134.8/km²
  (349/sq mi)
- ♕↑ 0.3%
- ⬡ 51.5 : 48.5%
- ♟♟ 92% Thai
  8% others
- ♟ Thai
- ▭ Buddhism
- ▦ Unitary
  parliamentary
  system under
  constitutional
  monarchy
- ⊕ ACD, ASEAN, G77,
  IMF, UN, WB, WTO
- ⬚ Thai Baht (THB)
- ▦ $406.84 bn
- ▦ $5,908
- ▦ Automotive industry,
  electronics,
  agriculture and food
  processing, tourism

LAOS

Mekong

Salween

MYANMAR

Ping

Chiang Rai

Phayao

Chiang Mai

Changwat Udon Thani

Phitsanulok

Khon Kaen

Mekong

Nakhon Sawan

Chao Phraya

Ubon Ratchathani

Phra Nakhon
Si Ayutthaya

Nakhon Ratchasima

BANGKOK
9.3 m

CAMBODIA

Chon Buri

Chanthaburi

*ANDAMAN
SEA*

Prachuap
Khiri Khan

Chumphon

*GULF OF
THAILAND*

Surat Thani

Nakhon Si Thammarat

Phuket

Phatthalung

Hat Yai

Pattani

Sadao

MALAYSIA

0 — 100 miles
0 — 100 km

Malaysia

Occupying the southern tip of the Malay Peninsula and the northern part of the island of Borneo, Malaysia comprises 13 states and 3 federal territories. British protectorates and colonies existed over the area during the 18th and 19th centuries. Nine of the states, all located in Peninsular Malaysia, originated as kingdoms and retained their royal

Malaysia
3.14° N, 101.69° E
UTC +8

☐ 330,800 km² (127,723 sq mi)	⚖ Federal parliamentary system under elective constitutional monarchy
O Tropical	
† 31,187,265	
⊡ 94.9/km² (37/sq mi)	⊕ ACD, ASEAN, CON, G-15, G77, IMF, NAM, UN, WB, WTO
↑↑ 1.5%	
◌ 75.4 : 24.6%	▯ Malaysian Ringgit (MYR)
⚲ 50% Malay 23% Chinese 12% indigenous 15% others	▤ $296.36 bn
	▦ $9,503
☻ Malay	▨ Electronics, chemicals, rubber, natural gas, tourism
▢ Islam	

THAILAND

George Town
⌂ • Kulim
• Taiping
• Ipoh
Sungai Siput

• Kota Baharu

Kuala Terenggan
Kuala Dungun

Kuantan

Kelang
★ KUALA
LUMPUR
6.8 m

MALACCA STRAIT

• Melaka

Jurong

INDONESIA

SINGAPORE

families. Together with the settlements of Penang and Malacca, they formed the Federation of Malaya in 1948, which gained independence in 1957. Six years later, they united with former British colonies of Sarawak and North Borneo; Singapore joined at the same time, but left the federation in 1965. Under a system of constitutional monarchy, the King of Malaysia is elected every five years by nine hereditary state rulers. However, state and national government are both largely in the hands of democratically elected officials. Independent Malaysia was initially highly reliant on exporting raw materials, particularly tin, rubber and palm oil. Since the 1980s, it has diversified into other sectors, such as manufacturing, banking and tourism, to become a prosperous, industrialized country.

PHILIPPINES

Kota Kinabalu

Sandakan

Beaufort

SOUTH CHINA SEA

Miri

BRUNEI

Tawau

Bintulu

Baram

BORNEO

Sibu

0   240 miles

Rajang

0   240 km

Kuching

INDONESIA

Laos

The predecessor to Laos was the Kingdom of Lan Xang Hom Khao, a unified state from 1354–1707 that splintered into three kingdoms. Absorbed from 1893 into the French colony of Indochina that also incorporated Vietnam and Cambodia, the three kingdoms were reunited into an autonomous state in 1949 and won full independence as a constitutional monarchy in 1953. However, the country was quickly torn apart by civil war between American-backed royalists and the communist Pathet Lao, who had Soviet and Chinese support. Communist victory forced the abdication of the king and established the current one-party socialist regime, which is closely allied to Vietnam.

From 1988, the Lao government began a series of liberal reforms allowing free enterprise and encouraging foreign investment. Important initiatives involve harnessing the country's rivers to generate hydroelectricity for domestic use and sale to neighbouring states, and the construction of a railway network connecting with regional neighbours.

Lao People's Democratic Republic
17.98° N, 102.63° E
UTC + 7

▢	236,800 km² 91,429 sq mi)	📖	Buddhism
○	Tropical monsoon	⚒	Unitary one-party socialist republic
✝	6,758,353		
▣	29.3/km² (76/sq mi)	⚖	ACD, ASEAN, G77, IMF, NAM, UN, WB, WTO
↑↑	1.4%		
▢	39.7 : 60.3%	💵	Lao Kip (LAK)
👥	53% Lao 11% Khmou 36% others	💹	$15.90 bn
		💰	$2,353
✠	Lao	⛏	Mining, timber, agriculture

Map labels:
CHINA
MYANMAR
VIETNAM
Phongsali
Louang Namtha
Xam Nua
Mekong
Louangphrabang
Ban Ban
Xiangkhoang
Muang Pakxan
THAILAND
VIENTIANE 997 K
THAILAND
Muang Khammouan
Savannakhet
SOUTH CHINA SEA
VIETNAM
Saravan
Pakxe
Ye Kong
Attapu
CAMBODIA

0 — 100 miles
0 — 100 km

Vietnam

Covering the east coast of mainland Southeast Asia, in 938 CE Vietnam threw off Chinese imperial dominion to become an independent kingdom. By 1887, the territory had become part of France's colonial possessions in Indochina. Excluding a brief period of wartime Japanese occupation, it remained so until 1954, when the pro-independence communist Viêt Minh forced France to withdraw. The country was then divided between communist north and pro-western south. Conflict between the two states led to an American-led foreign intervention to prop up the regime in the south, which escalated into a war that ended with northern victory in 1975. The country reunified under communist rule, but recovery from decades of war and division was slow and the country remained politically isolated. In 1986, the government began its *Doi Moi* (Renovation) reforms, liberalizing and industrializing the economy and opening the country to the rest of the world. Although Vietnam remains a one-party communist state, its government's embrace of capitalism has made it one of the fastest-growing economies in the world.

Socialist Republic of Vietnam
21.03° N, 105.83° E
UTC + 7

☐	330,967 km² (127,787 sq mi)
○	Tropical in south, monsoonal in north
♦	92,701,100
⊞	299/km² (775/sq mi)
⇅	1.1%
◻	34.2 : 65.8%
⋔	86% Vietnamese 14% others
♀	Vietnamese
⌑	Vietnamese folk religion
⚒	Unitary one-party socialist republic
⦿	ACD, ASEAN, G77, IMF, NAM, UN, WB, WTO
⬗	Vietnamese Dong (VND)
⊞	$202.62 bn
⧉	$2,186
⛴	Agriculture, clothing

Cambodia

The Khmer Empire, Cambodia's predecessor, was founded in the early ninth century and ruled much of Southeast Asia. Its capital, Angkor, was one of the largest cities in the medieval world and the site of an extensive temple complex, Angkor Wat, which appears on the national flag. Khmer power declined during the 15th century and the country became a vassal state dominated by neighbouring powers. During the 1860s, the Cambodian king placed the country under French protection and it became part of their empire. Independence was won in 1953, but in 1967 a civil war began between the communist Khmer Rouge and American-backed right-wing royalists. The Khmer Rouge won victory in 1975 and carried out a brutal genocide that killed at least 1.5 million people. Four years later, Vietnamese forces invaded and occupied the country, igniting another civil war that lasted until 1991. After a period under UN administration, the Kingdom of Cambodia was re-established as a constitutional monarchy in 1993. Despite some recent economic growth, the country remains wracked by authoritarian government, corruption and poverty.

THAILAND

LAOS

THAILAND

Sisophon
Siemreab
Lumphat
Stoeng Treng
Batdambang
Tonle' Sap
Mekong
Tonle Srepot
Pouthisat
Kampong Thurn
Kracheh
Krong Kaoh Kong
PHNOM PENH 1.7 m ★
Kampong Cham
VIETNAM
Kampong Spoe
Svay Rieng
GULF OF THAILAND
Kampot
HaTien
Mekong

☐	181,040 km² (69, sq mi)	
○	Tropical	
🕴	15,762,370	
🔢	89.3/km² (231/sq mi)	
↟↟	1.6%	
◻	20.9 : 79.1%	
🏘	98% Khmer 2% others	
🕴	Khmer	
🏛	Buddhism	
⚓	Unitary dominant-party parliamentary system under elective constitutional monarchy	
⚙	ACD, ASEAN, G77, IMF, NAM, UN, WB, WTO	
💷	Cambodian Riel (KHR)	
🔳	$20.02 bn	
🔳	$1,270	
🔳	Agriculture, clothing, tourism	

0   100 miles
0   100 km

Kingdom of Cambodia
11.54° N, 104.89° E
UTC + 7

Singapore

The city-state of Singapore is located at the southern point of the Malay Peninsula on the eastern entrance to the Strait of Malacca, the major channel linking the Pacific and Indian oceans. It was once the site of a Malay port, established in the 14th century but burned down by the Portuguese in 1613 and abandoned. In 1819, British statesman Stamford Raffles, recognizing the site's potential, signed a treaty with the area's Malay ruler that allowed him to establish a trading colony there. Singapore subsequently became a major port, and the population increased owing to migration from China and South Asia. British rule ended in 1963; initially the city joined the Malaysian Federation, but left two years later to become fully independent. The nation's modern founding father, Lee Kuan Yew, served as the country's first prime minister, remaining in office until 1990. His People's Action Party has won every election since independence. In spite of its small size, Singapore has become an economic powerhouse that serves as a global hub of transport and commerce.

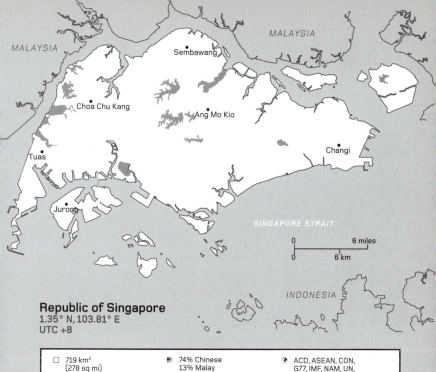

MALAYSIA

Sembawang

Choa Chu Kang

Ang Mo Kio

Changi

Tuas

Jurong

MALAYSIA

SINGAPORE STRAIT

0 ——— 6 miles
0 ——— 6 km

INDONESIA

Republic of Singapore
1.35° N, 103.81° E
UTC +8

□ 719 km² (278 sq mi)	⚥ 74% Chinese 13% Malay 13% others	☯ ACD, ASEAN, CON, G77, IMF, NAM, UN, WB, WTO
○ Tropical	☻ English, Malay, Mandarin, Tamil	⌨ Singapore Dollar (SGD)
⊡ 5,607,283	📖 Buddhism	▨ $296.97 bn
⊡ 7,908.7/km² (20,484/sq mi)	♨ Unitary dominant-party parliamentary republic	▦ $52,961
⇈ 1.3%		⛴ Financial services, electronics
◗ 100 : 0%		

Brunei

Situated on the northern coast of the island of Borneo, facing the South China Sea, the Sultanate of Brunei was established in 1368 and became the centre of a maritime empire from the 15th–17th centuries. The same family ruled from its foundation, but succession disputes and the encroachment of foreign colonial powers led to its territory and power diminishing over time. Brunei became a British protectorate in 1888, achieving domestic self-government in 1959 and full independence in 1984.

Today, the sultan has sweeping powers – technically, the country has been in a state of martial law since a rebellion in 1962. The sultan acts as both head of state and government, appointing the five councils that govern the country and controlling the national legislative council (according to the constitution, its members are to be chosen democratically, but only one election has ever been held, in 1962). Brunei's deposits of oil and natural gas, first discovered in 1929, provided a foundation for prosperity and economic growth that accelerated after independence.

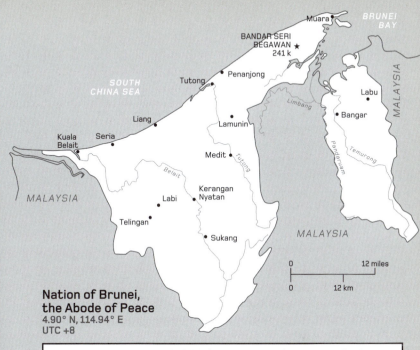

SOUTH
CHINA SEA

*BRUNEI
BAY*

MALAYSIA

Muara

BANDAR SERI
BEGAWAN
241 k

Labu

Penanjong

Bangar

Tutong

Limbang

Liang

Lamunin

Tutong

Pandaruam

Temurong

Kuala
Belait

Seria

Medit

Belait

Medit

Kerangan
Nyatan

Labi

Telingan

Sukang

MALAYSIA

MALAYSIA

0   12 miles

0   12 km

**Nation of Brunei,
the Abode of Peace**
4.90° N, 114.94° E
UTC +8

☐ 5,770 km² (2,228 sq mi)	⚥ 66% Malay 10% Chinese 24% others	🛒 Brunei Dollar (BND)
○ Tropical	♟ Malay, English	📖 $11.40 bn
⚑ 423,196	📖 Islam	📄 $26,939
⊞ 80.3/km² (208/sq mi)	⚒ Absolute monarchy	⛏ Oil and natural gas
⚥ 1.3%	◉ ACD, ASEAN, CON, G77, IMF, NAM, UN, WB, WTO	
◠ 77.5 : 22.5%		

Philippines

Located on an archipelago of over 7,100 islands, for most of its history the Philippines was split between local rulers. Spain colonized the area during the 16th century, renaming it in honour of their king, Philip II. Independence was declared in 1898, with the Philippine Republic established the following year. Self-rule was brief because the United States claimed possession of the islands, fighting a three-year war of conquest to solidify control. It wasn't until 1946 that the country gained independence again, having become a self-governing commonwealth in 1935. Achieving democracy was problematic, however; from 1965, the dictatorial and corrupt Ferdinand Marcos ruled as president until the nonviolent People Power Revolution of 1986 forced him to step down. The country has since transitioned to democratic rule, but has been troubled by corruption, debt, attempted coups, natural disasters and internal conflicts with communists and Muslim separatists in the south. Despite the country's many problems, the Philippines has a dynamic, industrializing economy that continued to grow after the 2008 global financial crisis.

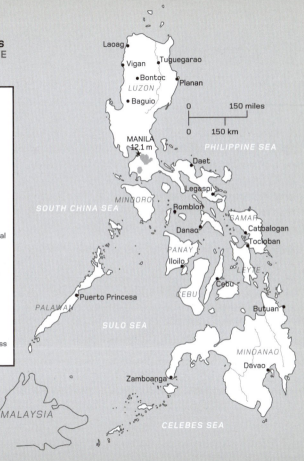

Republic of the Philippines
14.60° N, 120.98° E
UTC +8

- ☐ 300,000 km² (115,831 sq mi)
- ○ Tropical marine
- ✝ 103,320,222
- ▦ 346.5/km² (897/sq mi)
- ↑↑ 1.6%
- ◠ 44.3 : 55.7%
- ♞ 28% Tagalog 13% Cebuano 59% others
- ♞ Filipino
- 📖 Christianity
- ⚒ Unitary presidential constitutional republic
- ◉ ACD, ASEAN, G24, G77, IMF, NAM, UN, WB, WTO
- 🏷 Philippine Peso (PHP)
- ▦ $304.91 bn
- ▦ $2,951
- ▦ Electronics, aerospace, business outsourcing

Laoag
Vigan
Tuguegarao
Bontoc
Planan
Baguio
LUZON
MANILA
12.1 m
Daet
PHILIPPINE SEA
Legaspi
MINDORO
Romblon
SOUTH CHINA SEA
SAMAR
Catbalogan
Danao
Tocloban
PANAY
LEYTE
Iloilo
CEBU
Cebu
Butuan
PALAWAN
Puerto Princesa
SULO SEA
MINDANAO
Davao
Zamboanga
MALAYSIA
CELEBES SEA

0 — 150 miles
0 — 150 km

Timor-Leste

With steep, rugged, forested highlands and coastal lowlands fringed by reefs, Timor-Leste was initially settled some 42,000 years ago by people of South Asian origin. Melanesian and East Asian migrants joined them from 3000 BCE. The Portuguese arrived in the east of Timor island during the 16th century, declaring it the colony of Portuguese Timor in 1702; the Dutch established dominance over the western half (subsequently part of Indonesia). Independence was declared in 1975 but Indonesia invaded within a month, under the pretext that a communist state would be established. So began a period of oppressive occupation that led to the deaths of at least 100,000 people. In 1999, Indonesia held a referendum in which more than 75 per cent voted for independence. A three-year transition period under UN administration ensued, with sovereign status restored in 2002 under the Portuguese name, Timor-Leste. In 2006, internal violence led to the deployment of a UN peacekeeping mission that lasted until 2012. Since then the country has stabilized, although it remains one of the poorest in the region.

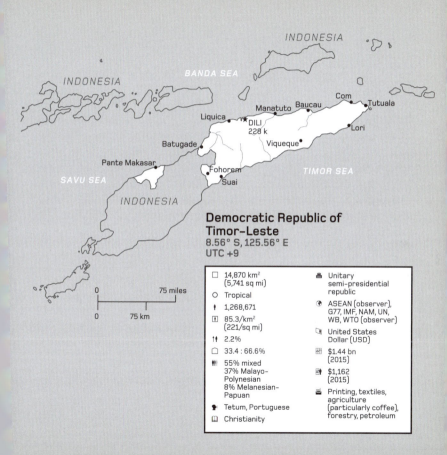

INDONESIA

BANDA SEA

INDONESIA

Com
Manatuto Baucau Tutuala
Liquica DILI Lori
228 k
Batugade Viqueque

Pante Makasar TIMOR SEA

SAVU SEA Fohorem
Suai

INDONESIA

Democratic Republic of Timor-Leste
8.56° S, 125.56° E
UTC +9

☐ 14,870 km² (5,741 sq mi)	🏛 Unitary semi-presidential republic
○ Tropical	
☦ 1,268,671	⊕ ASEAN (observer), G77, IMF, NAM, UN, WB, WTO (observer)
⊞ 85.3/km² (221/sq mi)	
⇅ 2.2%	💵 United States Dollar (USD)
◡ 33.4 : 66.6%	
🚻 55% mixed 37% Malayo-Polynesian 8% Melanesian-Papuan	📊 $1.44 bn (2015)
	📈 $1,162 (2015)
🗣 Tetum, Portuguese	🏭 Printing, textiles, agriculture (particularly coffee), forestry, petroleum
✝ Christianity	

0 ——— 75 miles

0 ——— 75 km

Democratic People's Republic of Korea

The 'hermit kingdom' of the Democratic People's Republic of Korea is perhaps the world's most culturally and politically isolated state. From 1392 to 1910, it was part of a kingdom that covered the entire Korean Peninsula. The Japanese then ruled until the end of the Second World War. A communist regime was established in the north of the peninsula in 1948. With Chinese and Soviet support, it invaded its southern neighbour in 1950, starting the three-year Korean War. The country's first president was Kim Il-sung, whose political ideology of *Juche*, which stresses self-reliance and patriotism, continues to be a major influence. Although he died in 1994, single-party totalitarian dictatorship has continued under his heirs. The collapse of the USSR cut off a major source of foreign support and contributed to a nationwide famine that lasted from 1994–98. The country remains underdeveloped, with a state-run economy and a regime committed to prioritizing military strength above all else. Consequently, the North Korean military is the world's fourth largest and, despite global condemnation, has developed nuclear capability.

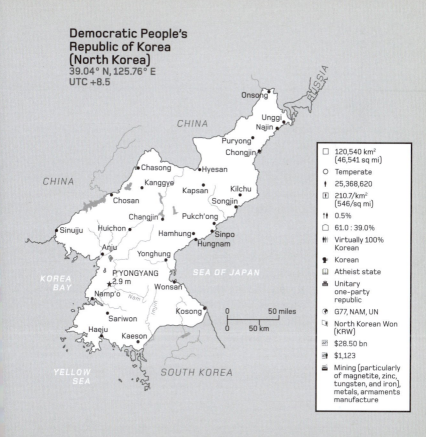

Democratic People's
Republic of Korea
(North Korea)
39.04° N, 125.76° E
UTC +8.5

CHINA

Onsong

Unggi
Najin

CHINA

RUSSIA

Puryong
Chongjin

Chasong

Hyesan

Kanggye

Kapsan

Kilchu

Chosan

Songjin

Changjin

Pukch'ong

Sinuiju

Huichon

Hamhung

Sinpo

Anju

Hungnam

Yonghung

P'YONGYANG
2.9 m

Wonsan

SEA OF JAPAN

Namp'o

Nam

Taedong

KOREA
BAY

Kosong

Sariwon

Haeju

Kaeson

YELLOW
SEA

SOUTH KOREA

☐	120,540 km² (46,541 sq mi)
O	Temperate
♦	25,368,620
⊞	210.7/km² (546/sq mi)
⇅	0.5%
◠	61.0 : 39.0%
♔	Virtually 100% Korean
☚	Korean
⛪	Atheist state
⚒	Unitary one-party republic
☻	G77, NAM, UN
⬡	North Korean Won (KRW)
⊞	$28.50 bn
▧	$1,123
⛏	Mining (particularly of magnetite, zinc, tungsten, and iron), metals, armaments manufacture

0   50 miles
0   50 km

Republic of Korea

In 1392, the Joseon dynasty established a kingdom that unified the Korean Peninsula, and which only came to an end in 1910 when Japan annexed the country, ruling until the end of the Second World War. Under the Allied partition plan, the peninsula was divided in two, with the democratic Republic of Korea established in the south in 1948. With the support of the UN, this country fought off a communist invasion from the north during the Korean War (1950–53). The postwar armistice created a demilitarized buffer zone that divided the two Koreas along the 38th parallel. Although at peace in the late 1950s, the Republic of Korea suffered from corrupt and autocratic government. In 1961, a military coup ushered in a period of unelected rulers that lasted until a pro-democracy movement swept the country in 1987, leading to a new democratic constitution and the accession of a directly elected president the following year. Before the 1960s, the country was relatively poor, but since then it has experienced rapid growth and become one of the world's most advanced states, largely thanks to an export-focused manufacturing sector.

Republic of Korea (South Korea)
37.57° N, 126.98° E
UTC +9

NORTH KOREA

Imjin

• Chorwon
• Yanggu
• Chunchon
• Kangyang
Inch'on •
★ SEOUL
9.9 m
• Samchok
• Chingju
Chongju •
• Andong
KOREA BAY
• Taejon
• Sangju
• Yondok
Kunsan •
• Kumch'on
• Puhang
• Chonju
• Taegu
YELLOW SEA
• Chinju
• Masan
• Kwangju
Mokp'o •
• Pusan

KOREA BAY

YELLOW SEA

SEA OF JAPAN

KOREAN STRAIT

CHEJU-DO

JAPAN

☐	100,280 km² (38,718 sq mi)
O	Temperate
✝	51,245,707
⊞	525.7/km² (1,362/sq mi)
✝✝	0.5%
☐	82.6 : 17.4%
✝✝✝	Virtually 100% Korean
✿	Korean
📖	Korean shamanism
⚖	Unitary presidential constitutional republic
✪	IMF, OECD, UN, WB, WTO
💱	South Korean Won (KRW)
▦	$1.41 tn
💰	$27,539
⚙	Electronics, telecommunications, automobile manufacture

0 50 miles
0 50 km

Japan

Made up of four main islands (Hokkaido, Honshu, Shikoku and Kyushu) and over 6,000 smaller ones, Japan's landscape is mostly mountainous and forested. Founded in 660 BCE, with the establishment of the imperial dynasty that still reigns, it is the world's oldest continuous monarchy. Imperial power was not always strong; from the 12th century, Japan was dominated by shoguns – military dictators who ruled in the emperor's name, and after 1603 initiated a policy of isolation. This ended in the mid-19th century; Japan opened up to Western influence and industrialized and modernized. Direct imperial rule was restored in 1868; over the next 70 years, Japan established a colonial empire that ruled Korea, Taiwan and parts of northeastern China. In 1941, Japan entered the Second World War, initially extending its empire across East Asia. Defeated in 1945, Japan underwent Allied occupation until 1952. The country became a constitutional monarchy with an elected parliament during this time. Despite heavy wartime losses, the economy experienced a dramatic resurgence, and the country is a major manufacturer and one of the world's largest economies.

Japan
35.69° N, 139.69° E
UTC +9

☐	372,962 km² (145,932 sq mi)	♨	Unitary parliamentary system under constitutional monarchy
○	Tropical in south to cool temperate in north	☀	ACD, G7, IMF, OECD, UN, WB, WTO
✝	126,994,511	☐	Japanese Yen (JPY)
⊡	348.4/km² (902/sq mi)	▩	$4.94 tn
♂♀	-0.1%	▩	$38,895
◖	94.0 : 6.0%	⊟	Motor vehicles, electronics manufacture
♟	98.5% Japanese 1.5% others		
☻	Japanese		
▣	Shinto		

RUSSIA

Wakkana

HOKKAIDO

•Asahikawa

Sapporo•

Kushiro

Aomori

PACIFIC OCEAN

Sendai

Niigata   Fukushima

HONSHU

SEA OF JAPAN

SOUTH KOREA

TOKYO
37.8 m ★

Matsue   Nagoya   Yokohama

Hiroshima   Kyoto•
Osaka

KOREAN STRAIT

KYUSHU   SHIKOKU

Sasebo•

Nagasaki•

Kagoshima

EAST CHINA SEA

PHILIPPINE SEA

0		200 miles
0		200 km

Australasia

Home to, perhaps, the oldest continuous civilization on Earth, Australasia also contains one of the last substantial landmasses to have been settled by humans. Whereas the Aboriginal people settled Australia at least 50,000 years ago, the Māori began to migrate to New Zealand around 700 years ago. Owing to their geographic isolation, Australia and New Zealand developed unique flora and fauna, such as the kangaroo and the kiwi. Their landscapes and cultures differ widely, however: New Zealand is mainly temperate, while Australia varies from desert to tropical rainforest. The trajectories of the two areas converged during the 17th century, when separate Dutch expeditions were the first Europeans to sight them. From the late 18th century, European migrants began to arrive in both places, establishing colonial regimes, after which the indigenous populations declined rapidly. Both colonies achieved self-government by the 1850s and independence by the 1940s. Australasia is stable and highly developed, although its prosperity has not always been equitably shared with its first peoples.

Australia

The Australian landmass, together with Tasmania, lies between the Pacific and Indian Oceans and has a landscape of deserts, rainforests and mountains. The indigenous Australians, known broadly as Aboriginals despite having regionally heterogeneous cultures and languages, arrived from Asia at least 50,000 years ago. In 1788, the British settled Australia, establishing a penal colony – New South Wales. Over subsequent decades, migration from the British Isles escalated and other colonies emerged across Australia, while the Aboriginal population was decimated through conflict with the settlers and Western diseases. The colonies federated in 1901 and became independent from Britain in 1942; the six states retained their own constitutions, laws and governments (two additional mainland territories have slightly less autonomy). From the mid-1800s, the country has became increasingly culturally diverse owing to migrants from continental Europe and, more recently, Asia. Since the gold rush of the 1850s, mining has been an important part of the economy; Australia is the world's largest exporter of many commodities, including coal and iron ore.

INDONESIA

ARAFURA SEA

PAPUA NEW GUINEA

TIMOR SEA

• Darwin

• Wyndham

• Derby

Ashburton

GREAT DIVIDING RANGE

• Cooktown

• Cairns

CORAL SEA

• Townsville

• Cloncurry

• Alice Springs

• Rockhampton

• Charleville

• Geraldton

• Coolgardie

• Perth

• Port Augusta

• Bourke

Darling

• Brisbane

• Newcastle

• Sydney
4.5 m

TASMAN SEA

GREAT AUSTRALIAN BIGHT

• Adelaide

Murray

★ CANBERRA
423 k

• Albany

• Melbourne

TASMANIA

• Hobart

| | 500 miles |
| 0 | 500 km |

☐ 7,741,220 km²
(2,988,902 sq mi)

○ Mostly arid to semi-arid

† 24,127,159

▦ 3.1/km²
(8/sq mi)

†† 1.4%

⬜ 89.6 : 10.4%

♦♦ 92% European
8% other

☻ English

☐ Christianity

♟ Federal parliamentary system under constitutional monarchy

◉ CON, IMF, OECD, PIF, UN, WB, WTO

☒ Australian Dollar (AUD)

▦ $1.20 tn

▦ $49,928

▦ Mining (particularly coal, gold and iron-ore), agriculture and food processing

Christmas Island, Cocos (Keeling) Islands and Norfolk Island are all under Australian administration. Australia also claims sovereignty over Australian Antarctic Territory.

Commonwealth of Australia
35.28° S, 149.13° E
UTC +8 to UTC +10.5

New Zealand

Consisting of two main islands and 600 smaller ones, New Zealand sits at the confluence of the Tasman Sea and Pacific Ocean, 1,600 km (1,000 miles) from the nearest major landmass. South Island is largely mountainous with fjords and glaciers, while North Island has more arable land and is home to 75 per cent of the population. The Polynesian Māori people settled here in the late 13th century, making it one of the last areas to be populated by humans. The first European explorers to reach the country were the Dutch, in 1642. Full-scale, mostly British, colonization began in the 19th century, with the British Crown declaring sovereignty under the Treaty of Waitangi (1840). The Māori were devastated by warfare and Western diseases, and the majority of their land was redistributed to new settlers. During the 20th century, the flow of migrants diversified to other Europeans, Asians and Polynesians. New Zealand achieved self-governance in 1846, domestic independence in 1907 and full sovereign status in 1947. It has gone on to be a highly stable nation, and has the distinction of being among the least corrupt countries in the world.

**New Zealand
or Aotearoa**
41.29° S, 174.78° E
UTC +12

- □ 267,716 km² (103,363 sq mi)
- ○ Temperate
- ♦ 4,692,700
- ⊡ 17.8/km² (46/sq mi)
- ⇅ 2.1%
- ◖ 86.3 : 13.7%
- ⋔ 70% European
  14% Maori
  16% others
- ♠ English, Maori
- ▥ Christianity
- ♣ Unitary parliamentary system under constitutional monarchy
- ◉ CON, OECD, IMF, PIF, UN, WB, WTO
- ▧ New Zealand Dollar (NZD)
- ▨ $185.01 bn
- ▤ $39,427
- ▦ Agriculture and food processing, forestry, mining (particularly coal)

Auckland
1.3 m

Paeroa
Hamilton
Rotorua
Taupo
Gisborne

New Plymouth
NORTH ISLAND
Napier

Palmerston North

TASMAN SEA

Nelson
Blenheim
Westport
WELLINGTON
383 k

Greymouth
SOUTH ISLAND

PACIFIC OCEAN

Christchurch

Timaru

Oamaru

Dunedin

Invercargill

| 0 | 100 miles |
| 0 | 100 km |

The Cook Islands, Niue and Tokelau, have self-government but foreign policy that is mostly the responsibility of New Zealand. The Ross Dependency in Antarctica is also claimed by New Zealand.

Oceania

Totalling some 10,000 islands spread across 8.5 million km²
(3.3 million square miles) of the Pacific, Oceania is divided into
three geographical and cultural regions: Melanesia, Micronesia
and Polynesia. Human population had begun by 60,000 BCE, with the
Papuan peoples settling New Guinea. Around 4,000 years ago, they
interacted with Austronesian migrants from Southeast Asia, and
ventured further east into Melanesia. By 1000 BCE, Micronesia, to
the north, was settled; a final wave of migration began 2,000 years
ago and moved eastwards to Polynesia. Europeans first explored
the continent in the 16th century, but it was not until the 19th
that Western imperialist activity became widespread. Indigenous
populations, often undermined by foreign disease, were placed
under colonial government; the islands were valued for their
natural resources as well as their strategic importance as transit
and supply points for shipping. The majority of Oceania gained
independence in the late 20th century. The islands remain reliant
on tourism and the primary sector; they are also vulnerable to
changing environmental conditions, particularly rising sea levels.

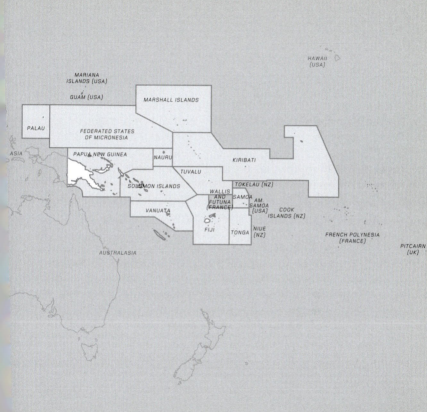

Palau

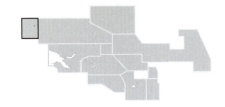

This archipelago in the western Pacific consists of eight main islands and over 300 smaller ones, the vast majority of which form a lagoon ringed by a barrier reef. Palau was first settled around 3,000 years ago by migrants from Indonesia and, subsequently, from New Guinea, the Philippines and Polynesia. The country was part of the Spanish empire for over 300 years, until Spain sold it to Germany in 1899. Japan annexed Palau in 1914, ruling until 1944. Following the Second World War, a UN mandate placed Palau under American administration. The country voted against joining the Federated States of Micronesia in 1979, opting instead to seek self-government. Independence was achieved in 1994, although Palau retains close ties with the United States. Under the terms of the Compact of Free Association, the US government provides Palau with financial assistance, as well as being obliged to defend it. Thanks to its fisheries, Palau is one of the most prosperous states in Oceania. In 2009, it created the world's first shark sanctuary, setting aside 600,000 km^2 (232,000 square miles) of its waters for them.

Republic of Palau
7.50° N, 134.62° E
UTC +9

☐	460 km² (178 sq mi)
○	Tropical
✝	21,503
⊞	46.7/km² (121/sq mi)
⇅	1.0%
◠	87.6 : 12.4%
⋔	73% Palauan 22% Asian 5% others
✒	English, Palauan
◻	Christianity
⛏	Unitary presidential constitutional republic
◉	IMF, PIF, UN, WB
▭	United States Dollar (USD)
▦	$293.0 m
▤	$13,626
⛏	Agriculture, fishing, tourism

• Agol

NGERULMUD ★
400

• Ngatpang

Koror City
14 k

Alrai •

PHILIPPINE SEA

KOROR

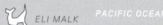

NIGERUKTABEL

PACIFIC OCEAN

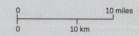

ELI MALK

0	10 miles
0	10 km

BELILIOU

Salpan

ANGAUR

Federated States of Micronesia

Made up of 607 islands in the western South Pacific, the FSM spreads some 2,700 km (1,680 miles) from west to east. As a result of its far-flung landmass, the FSM has economic rights over waters covering an area of four million km² (1.55 million sq mi). The islands were first settled by the Micronesian peoples in around 2,000 BCE. From the 1880s to the 1940s, they were successively part of the Spanish, German and Japanese empires. After the Second World War, the country was placed under American

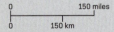

ULITHI ATOLL
Colonia
YAP
FAIS

PACIFIC OCEAN

NAMOUNITO ATO

NGULU
ATOLL

0   150 miles

0   150 km

administration as part of a UN trusteeship. In 1979, the four states that comprise the FSM (Yap, Chuuk, Pohnpei and Kosrae – each with its own distinct culture) ratified a new constitution under which they retained considerable political autonomy. This paved the way to full independence in 1986, although the FSM retained strong economic and political ties with the United States under the Compact of Free Association. Today, the FSM is reliant on American financial support, and faces long-term environmental problems owing to the impact of climate change and overfishing.

□ 700 km² (270 sq mi)	49% Chuukese 30% Pohnpeian 21% others	G77, IMF, PIF, UN, WB
O Tropical	English (as well as seventeen indigenous languages)	United States Dollar (USD)
104,937		$322.0 m
149.9/km² (388/sq mi)	Christianity	$3,069
0.5%	Federal parliamentary republic	Fishing
22.5 : 77.5%		

HALL ISLANDS

Federated States of Micronesia (FSM)
6.91° N, 158.16° E
UTC +10 to UTC +11

WENO
14 K

ORULUK ATOLL

PALIKIR
6 K

PACIFIC OCEAN

CHUUK

POHNPEI

KUKUNOR ATOLL

MAJURO

Papua
New Guinea

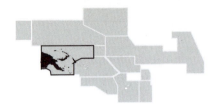

Ethnically and linguistically heterogeneous, Papua New Guinea (PNG) is home to over 800 indigenous languages. Covering the eastern half of New Guinea (the west is part of Indonesia) and dozens of offshore islands, its landscape features rainforests, mountains, swamps and grassland. First settled more than 40,000 years ago by the Papuan people, it was colonized in the late 19th century; Germany ruled to the north and Britain to the south. In 1906, Britain transferred control of its territory to Australia, which also administered the former German area after the First World War. Australian rule continued until PNG became independent in 1975. Building national unity was hampered by Bougainville Island's attempt to become independent. Although it rejoined PNG in 1976, a secessionist war from 1988–98 ended with Bougainville becoming an autonomous region, with the promise of a referendum on independence by 2020. PNG has huge potential wealth owing to mineral deposits; however, inhospitable terrain and lack of infrastructure makes accessing them problematic. As much as 85 per cent of the population is engaged in subsistence agriculture.

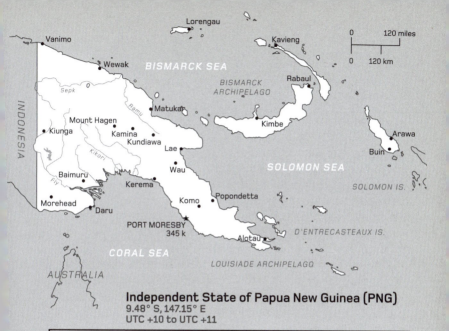

Independent State of Papua New Guinea (PNG)
9.48° S, 147.15° E
UTC +10 to UTC +11

☐ 462,840 km² (178,704 sq mi)	✤ Hiri Motu, Tok Pisin, English	▭ Papua New Guinean Kina (PGK)
○ Tropical	▭ Christianity	▦ $16.93 bn (2014)
✝ 8,084,991	♨ Unitary parliamentary system under constitutional monarchy	▤ $2,183 (2014)
▯ 17.9/km² (46/sq mi)		▬ Agriculture, forestry, mining
✝✝ 2.1%		
◻ 13.0 : 87.0%	⦿ ASEAN (observer), CON, G77, IMF, NAM, PIF, UN, WB, WTO	
♛ 99% Papuan and Melanesian 1% others		

Map labels: Lorengau, Vanimo, Kavieng, Wewak, BISMARCK SEA, BISMARCK ARCHIPELAGO, Rabaul, INDONESIA, Sepk, Ramu, Matukar, Kimbe, Mount Hagen, Kiunga, Kamina, Kundiawa, Kikori, Arawa, Buin, Lae, Wau, SOLOMON SEA, Baimuru, Fly, Kerema, Komo, Popondetta, SOLOMON IS., Morehead, Daru, PORT MORESBY 345 k, Alotau, D'ENTRECASTEAUX IS., CORAL SEA, LOUISIADE ARCHIPELAGO, AUSTRALIA

Solomon Islands

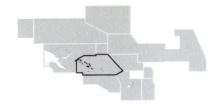

The indigenous Melanesians first arrived over 30,000 years ago, to settle the Solomon Islands' two parallel chains of volcanic islands and coral atolls. When the first Europeans arrived – a Spanish expedition in 1568 – they found gold; their leader claimed to have found the origin of King Solomon's wealth, hence the islands' name. Europeans did not return until the late 18th century, and during the 1890s the islands were placed under a British protectorate. Japan invaded in 1942, provoking a three-year Allied counter-attack that took place in dense tropical rainforest and saw some of the most arduous fighting of the Second World War. Years of campaigning brought independence in 1978, but ethnic rivalry has plagued the country since that time. In 1999, fighting began on the main island, Guadalcanal, between the local population and migrants from the neighbouring island of Malaita. Peace was made in 2000, but the conflict was highly disruptive, causing significant economic damage and creating lasting enmity. An international assistance mission deployed between 2003 and 2017 helped to restore stability to the country.

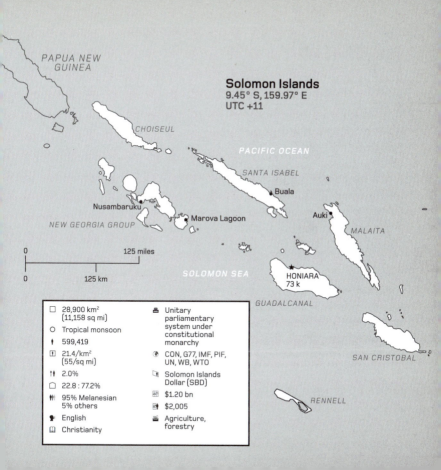

PAPUA NEW GUINEA

Solomon Islands
9.45° S, 159.97° E
UTC +11

CHOISEUL

PACIFIC OCEAN

SANTA ISABEL

Buala

Nusambaruku

Marova Lagoon

Auki

MALAITA

NEW GEORGIA GROUP

SOLOMON SEA

HONIARA
73 k

GUADALCANAL

SAN CRISTOBAL

RENNELL

| 125 miles | | |
| 125 km | | |

- ☐ 28,900 km² (11,158 sq mi)
- ○ Tropical monsoon
- ♀ 599,419
- ⊡ 21.4/km² (55/sq mi)
- ⇅ 2.0%
- ◻ 22.8 : 77.2%
- ♟ 95% Melanesian 5% others
- ♟ English
- ▢ Christianity

- ♨ Unitary parliamentary system under constitutional monarchy
- ☉ CON, G77, IMF, PIF, UN, WB, WTO
- ▢ Solomon Islands Dollar (SBD)
- ▦ $1.20 bn
- ▨ $2,005
- ▤ Agriculture, forestry

Marshall Islands

Stretching 1,300 km (800 miles) northwest to southeast, the Marshall Islands comprise 29 coral atolls, five islands and over 1,200 islets that form two parallel chains – Ratak (sunrise) in the east and Ralik (sunset) in the west. They were first settled some 3,000 years ago, by Micronesian peoples. Although Spain claimed the islands in the early 16th century, they are named after an Englishman, John Marshall, who explored the area in 1788. Germany bought the islands from Spain in 1885.

Japan seized the islands in 1914 during the opening weeks of the First World War, and continued to rule until 1944, when the United States invaded and occupied. The US regime continued after the war as part of a UN trusteeship; from 1946–58, it evacuated the Bikini and Enewetak atolls to use them for nuclear testing. The country won independence as a parliamentary democracy in 1986. Under the Compact of Free Association, the US is responsible for defence and provides aid, but also has the right to rent Kwajalein Atoll as a missile-testing site and military base.

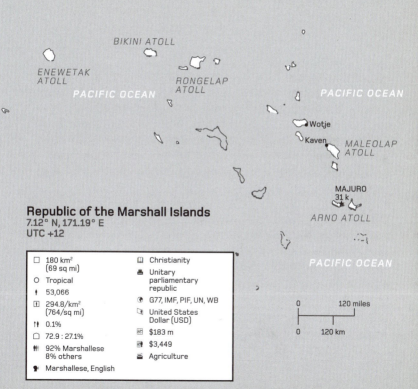

BIKINI ATOLL

ENEWETAK
ATOLL

RONGELAP
ATOLL

PACIFIC OCEAN

PACIFIC OCEAN

Wotje

Kaven

MALEOLAP
ATOLL

MAJURO
31 k

ARNO ATOLL

PACIFIC OCEAN

Republic of the Marshall Islands
7.12° N, 171.19° E
UTC +12

☐ 180 km² (69 sq mi)	Christianity
○ Tropical	Unitary parliamentary republic
✝ 53,066	G77, IMF, PIF, UN, WB
⬆ 294.8/km² (764/sq mi)	United States Dollar (USD)
⬆⬆ 0.1%	$183 m
◠ 72.9 : 27.1%	$3,449
⚣ 92% Marshallese 8% others	Agriculture
⚑ Marshallese, English	

0	120 miles
0	120 km

Vanuatu

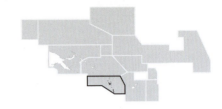

An archipelago of over 80 islands, around four-fifths of which are inhabited, Vanuatu is susceptible to tectonic activity, and has nine active volcanoes (two of which are undersea). Human settlement goes back to 2000 BCE, with the arrival of Melanesian settlers. Successive waves of migrants – mostly Melanesians and some Polynesians – have led to over 100 languages being spoken. Portuguese explorers landed on the largest island, which they called Espiritu Santo, in 1606. Europeans only returned to the islands in the late 18th century, naming them the New Hebrides. Western missionaries and traders arrived during the 19th century and, in an inhumane process known as 'black-birding', kidnapped thousands of indigenous people and forced them to work in plantations in Fiji, New Caledonia and Australia. In 1906, the islands were placed under Anglo-French rule that lasted until they won independence as Vanuatu in 1980. Despite accusations of corruption, the country has been largely peaceful; the most damaging event in Vanuatu's recent history occurred in 2015, when Cyclone Pam caused extensive damage.

Republic of Vanuatu
17.73° S, 168.33° E
UTC +11

▢	12,190 km² (4,707 sq mi)
○	Tropical
✝	270,402
▦	22.2/km² (57/sq mi)
✝✝	2.2%
▢	26.4 : 73.6%
✟	98% Ni-Vanuatu 2% others
✎	Bislama, French, English
▢	Christianity
▲	Unitary parliamentary republic
◉	CON, G24, G77, IMF, NAM, PIF, UN, WB, WTO
▭	Vanuatu Vatu (VUV)
▦	$773.5 m
▨	$2,861
▤	Agriculture, tourism

BANKS ISLANDS

Sola

ESPIRITU SANTO

Luganville

CORAL SEA

Panngi

PACIFIC OCEAN

MALAKULA

EFATE

PORT VILA
51 k

ERROMANGO

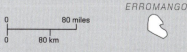

TANNA

Isangel

Nauru

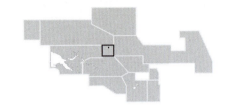

An oval-shaped island surrounded by coral reef, Nauru was once covered by phosphate minerals formed by centuries of accumulated bird droppings. The indigenous population descend from Micronesians, Polynesians and Melanesians, who began to arrive around 2,000 years ago. The island was first sighted by an English vessel in 1798, and Europeans began to arrive in the 1830s, with Germany annexing the island in 1888. The island's future was transformed when its phosphate deposits – then in high demand as a fertilizer – were discovered in 1900. Australian troops seized Nauru in 1914, and ruled it in a joint mandate with Britain and New Zealand that lasted until Japan occupied the island in 1942. After the Second World War, an Australian administration was instituted, and continued until Nauruan independence in 1968. As its phosphate dwindled in the 1990s, the country descended into economic crisis. In 2001, the country was forced to make a deal with Australia to house a detention centre for asylum seekers, despite attracting international condemnation for the conditions there.

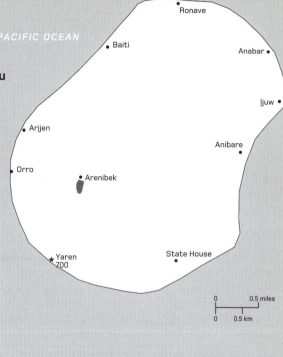

PACIFIC OCEAN

Republic of Nauru
0.55° S, 166.92° E
UTC +12

Ronave

Baiti

Anabar

Ijuw

Arijen

Anibare

Orro

Arenibek

Yaren
700

State House

☐	20 km² (8 sq mi)
○	Tropical
⚥	13,049
⊞	652.5/km² (1,690/sq mi)
⚤	4.5%
⌂	100 : 0%
⚤	58% Nauruan 26% other Pacific Islander 16% others
☻	Nauruan, English
▣	Christianity
♟	Parliamentary republic
◉	CON, G77, IMF, PIF, UN, WB
◖	Australian Dollar (AUD)
▤	$102.06 m
▥	$7,821
⛏	Phosphate mining, fishing licences, hosting an Australian detention centre

0   0.5 miles

0   0.5 km

Kiribati

Spread out over the Central Pacific, Kiribati totals 33 islands arranged in three groups: the Gilbert Islands, home to 90 per cent of the population, and the more sparsely populated Phoenix and Line Islands. The country also has a remote expanse of coral reefs untouched by humans. The Gilberts were settled by Micronesians in 3000 BCE, later joined by Samoans from the 11th–14th centuries. They became a British protectorate in 1892, which extended to include the Line Islands in 1919 and the Phoenix Islands in 1937, all of which were uninhabited. With the exception of a Japanese occupation from 1942–43, British rule continued until the country was made independent in 1979. By this time its once-rich phosphate deposits had been exhausted, and the resource-poor country's GDP per capita had dropped to the lowest in Oceania. As the majority of Kiribati's territory is composed of low-lying coral atolls, the country is vulnerable to rising ocean levels. The threat is so serious that, in 2014, the government purchased land in Fiji on which to relocate the population should the country become uninhabitable.

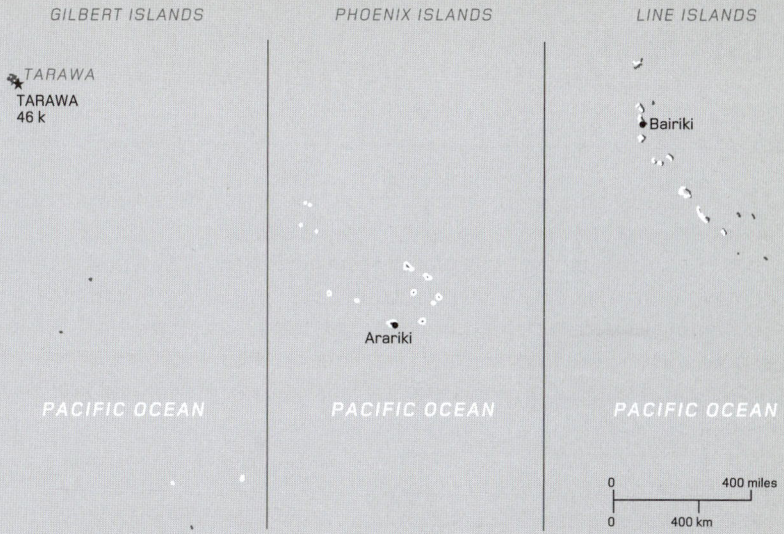

GILBERT ISLANDS   PHOENIX ISLANDS   LINE ISLANDS

TARAWA
TARAWA
46 k

Bairiki

Arariki

PACIFIC OCEAN   PACIFIC OCEAN   PACIFIC OCEAN

0   400 miles

0   400 km

Total east-west span of
Kiribati is approx 2150 km
(1335 miles)

Republic of Kiribati
1.45° N, 172.97° E
UTC +12 to UTC +14

☐ 810 km² (313 sq mi)	♟ 90% I-Kiribati 10% others	🖃 Kiribati Dollar (KID) (pegged to the Australian Dollar)
O Tropical	♟ English, Gilbertese	💹 $165.77 m
♟ 114,395	📖 Christianity	💴 $1,449
⊞ 141.2/km² (366/sq mi)	♟ Parliamentary republic	🎣 Fishing licences
♟♟ 1.8%	♟ CON, G77, IMF, PIF, UN, WB	
◠ 44.4 : 55.6%		

Tuvalu

A chain of nine Pacific islands, Tuvalu is low-lying – its highest peak, on the island of Niulakita, stands just 4.8 m (16 ft) above sea level. The first of several waves of Polynesian voyagers arrived in around 2000 BCE, and Spanish sailors sighted the islands in 1568. However, it was not until the early 19th century that Europeans began to visit regularly, naming the islands after British merchant and politician Edward Ellice. Due to foreign diseases and widespread abductions to overseas plantations, the population declined from 20,000 in 1850 to 3,000 in 1875. Britain established a protectorate over the Ellice Islands in 1892, which in 1916 were joined with the Gilbert Islands. Racial differences and economic rivalry led to demands for secession by the Ellice Islands, approved in a 1974 referendum. The next year they split off to form a separate colony named Tuvalu (meaning 'eight standing together' in reference to the number of inhabited islands). Tuvalu won independence in 1978. With few natural resources, the country has turned to a new source of income: their national domain name, ".tv", provides a major source of government revenue.

NANUMEA Lolua

NANUMANGA

NIUTAO
Kulia

Tonga

PACIFIC
OCEAN

NUI
Tanrake

VAITUPU
Asau

PACIFIC
OCEAN

NUKUFETAU

Savave

FUNAFUTI

★ FUNAFUTI
6 k

NUKULAELAE
Fangaua

NIULAKITA

□ 30 km²
(12 sq mi)

○ Tropical

♦ 11,097

⊞ 369.9/km²
(958/sq mi)

↟↟ 0.9%

◠ 60.6 : 39.4%

♯♯ 96% Polynesian
4% Micronesian

♟ Tuvaluan, English

▥ Christianity

♟ Non-partisan
parliamentary
democracy under
constitutional
monarchy

◉ CON, IMF, PIF, UN,
WB

◖ Tuvaluan Dollar
(TVD), linked to
Australian Dollar

▦ $34.22 m

▦ $3,084

▥ Fishing and fishing
licences, Internet
domain licensing

Tuvalu
8.52° S, 179.20° E
UTC +12

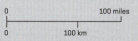

0		100 miles
0	100 km	

Fiji

Fiji is an archipelago of over 800 volcanic islands spread over 3 million km² (1.2 million square miles) in the South Pacific. Viti Levu, the largest, houses the capital and over two-thirds of the population. The majority of Fijians descend from Melanesian people who arrived over 3,500 years ago. Frequent warfare between island chiefs prevented a paramount ruler emerging until 1871. European missionaries and settlers were well established by that time and, in 1874, Fiji became a British colony. From 1879 to 1916, the British imported Indian indentured labourers to work on sugar plantations; many remained, and by the time independence was granted in 1970, they made up just under half of the population. In 1987, an Indo-Fijian politician won the elections, but was overthrown when military coups re-established indigenous rule. Another Indo-Fijian won power in 1999, but was ejected the following year. This political turmoil and discrimination led to thousands of Indo-Fijians fleeing the country. Since 2000, the military has been a major force in Fijian politics; it seized power in a 2006 coup and its candidate won elections held in 2014.

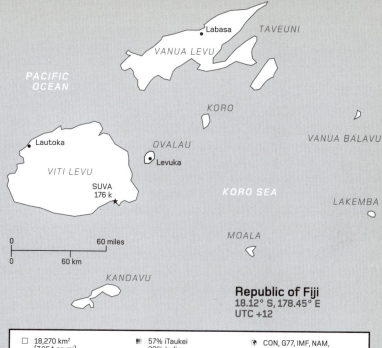

PACIFIC
OCEAN

VANUA LEVU

Labasa

TAVEUNI

KORO

VANUA BALAVU

Lautoka

OVALAU

Levuka

VITI LEVU

SUVA
176 k

KORO SEA

LAKEMBA

MOALA

KANDAVU

0 ————— 60 miles
0 ————— 60 km

Republic of Fiji
18.12° S, 178.45° E
UTC +12

☐ 18,270 km² (7,054 sq mi)	🏃 57% iTaukei 38% Indian 5% others	🌐 CON, G77, IMF, NAM, PIF, UN, WB, WTO
○ Tropical marine	🗣 English, Fijian, Hindi	💱 Fijian Dollar (FJD)
⚲ 898,760	📖 Christianity, Hinduism	💵 $4.63 bn
⊞ 49.2/km² (127/sq mi)	⚖ Unitary parliamentary constitutional republic	💴 $5,153
⇅ 0.7%		🏛 Tourism, agriculture
⌂ 54.1 : 45.9%		

Samoa

The Samoan archipelago consists of 15 islands settled by Polynesian migrants around 1000 BCE. Europeans first explored in 1722 and Christian missionaries arrived in 1830. Owing to the islands' strategic position in the central South Pacific, Western powers vied for dominance. The islands were unilaterally annexed in 1899; the United States gained the six eastern islands (now American Samoa), while Germany took the nine islands in the west.

German rule of Western Samoa continued until 1914, when New Zealand forces occupied the territory during the First World War. They administered the territory until its independence in 1962 – the first island-state in Oceania to be freed from colonial rule. Hereditary chiefs called *matai* are highly influential, dominating political life. Two major changes occurred in the 1990s: universal suffrage was adopted in 1990, and seven years later the country officially dropped 'Western' from its name. Samoa's economy is largely dependent on the food production and foreign remittances, and is highly vulnerable to cyclones and tsunamis.

Independent State of Samoa
13.85° S, 171.75° W
UTC +13

☐ 2,840 km² (1,097 sq mi)	♔ 93% Samoan 7% others	▢ Samoan Tala (WST)
○ Tropical	♟ Samoan, English	▦ $785.91 m
♱ 195,125	▢ Christianity	▨ $4,028
⊞ 68.9/km² (178/sq mi)	♨ Unitary parliamentary democracy	⛴ Agriculture, fishing, tourism
♰ 0.7%		
◲ 19.0 : 81.0%	⦿ CON, G77, IMF, PIF, UN, WB, WTO	

Tonga

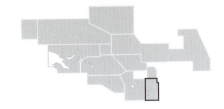

A group of 170 islands stretching 800 km (500 miles) from north to south, Tonga is home to the last reigning Polynesian monarch. It sits on the Tonga Trench, a depression in the ocean floor that includes Horizon Deep, Earth's second-deepest point at 10,881 m (35,700 ft) below sea level. First peopled by Polynesian settlers around 3,000 years ago, the islands were united under a paramount ruler by the tenth century. From 1200 to 1500, Tonga was the most powerful force in the Central Pacific. Sustained contact with Europeans began in the 1770s with the arrival of Captain Cook, who called the country 'The Friendly Islands'. Christian missionaries arrived in 1797 and converted the king in 1831, after which the new faith spread rapidly. A constitutional monarchy was declared in 1875, and although the country became a British protectorate in 1900, it largely maintained domestic self-rule. Tonga became independent in 1970, and its king continues to be influential, despite having relinquished many of his executive powers in 2008. The economy is largely dominated by agriculture, but is reliant on imports of food as well as manufactured goods and remittances from abroad.

○ NIUAFO'OU

TAFAHI
○
NIUATOPUTAPU

Kingdom of Tonga
21.14° S, 175.20° W
UTC +13

SOUTH
PACIFIC
OCEAN

☐	750 km² (290 sq mi)
○	Tropical
✝	107,122
⊞	148.8/km² (385/sq mi)
↕	0.7%
◖	23.8 : 76.2%
♈	97% Tongan 3% others
☘	Tongan, English
▢	Christianity
♣	Unitary parliamentary system under constitutional monarchy
◉	CON, G77, IMF, PIF, UN, WB, WTO
⛶	Tongan Pa'anga (TOP)
▦	$395.16 m
▤	$3,689
⛴	Agriculture, tourism

FONUALEI
○

VAVA'U
Neiafu ●
LATE ○

PACIFIC
OCEAN

TOFUA ○   ● Pangai
LITUKA

NOMUKA ○

0		100 miles
0	100 km	

NUKU'ALOFA
25 k
TONGATAPU
OHONUA

North America

The vast majority of the North American landmass and its population is contained in Canada, the United States and Mexico. The rest of the continent comprises the seven nations of Central America and the island-states of the Caribbean. Geographically, North America's dominant features are the Rocky Mountains that dominate its western half, and the extensive steppe plains (prairies) to the east. The Rockies have such an effect on geography that nearly all of the continent's major watersheds drain east towards the Atlantic, or northeast into the Great Lakes region. Much of the continent's northern half bears signs of extensive scouring by glaciers during the last Ice Age – the period when humans first arrived via a land bridge across the Bering Strait from Asia. Although the exact sequence remains controversial, human settlement is thought to have begun at least 14,000 years ago. Subsequently, groups of 'Paleo-Indian' settlers developed into the Native American tribes of North America and the 'Meso-American' cultures of Mexico and Central America, prior to the arrival of European colonists from the 16th century.

United States of America

With environments ranging from Arctic tundra to tropical rainforest, and from arid desert to rolling plains, the United States (US) is richly endowed with natural resources that include vast deposits of oil and natural gas. Native Americans settled here 14,000 years ago, and European colonization began in the 16th century. In 1776, 13 British colonies declared independence, instituting a federal system of government with an elected president holding executive power, although individual states retained some autonomy. After the Revolutionary War that secured independence, the country expanded westwards, increasing the number of states (there are now 50). The economy and population grew rapidly, stimulated by an influx of migrants from around the world. Until the mid-19th century, slaves originating from Africa played a major role in the economy, particularly in the South. Tensions between North and South led to a Civil War (1861–65), which ultimately ended slavery while preserving the country's unity. The US played a major role in Allied victories during both world wars, and has remained the leading Western power since the 1940s.

United States of America (USA)

38.90° N, 77.03° W
UTC −4 to UTC −8, UTC −10, UTC −11

- □ 9,831,510 km² (3,795,967 sq mi)
- ○ Mostly temperate
- ⚤ 323,127,513
- ⊞ 35.3/km² (91/sq mi)
- ⚤ 0.7%
- ⬚ 81.8 : 18.2%
- ⚤ 72% white
  13% black
  15% others
- ✿ English, Spanish, Hawaiian, plus indigenous languages
- ▯ Christianity

- ♜ Federal presidential constitutional republic
- ◉ G7, IMF, NAFTA, NATO, OAS, OECD, UN, WB, WTO
- ▭ United States Dollar (USD)
- ▥ $18.57 tn
- ▤ $57,467
- ⚒ Manufacturing, telecommunications, pharmaceuticals and medical equipment, military equipment, agriculture and food processing

American Samoa, Guam, Northern Mariana Islands, Puerto Rico and the US Virgin Islands are all under control of the US federal government, but are not part of the United States of America. They have domestic self-government and elect their own governors but are not states.

RUSSIA
Pt. Barrow
ALASKA · Fairbanks
· Anchorage
CANADA

CANADA

PACIFIC OCEAN

Honolulu
HAWAII

Seattle
Olympia
Salem
Boise
Helena
Bismarck
St. Paul
Pierre
Madison
Detroit
Montpelier
Concord
Albany
Boston 8.5 m
New York
Sacramento
Carson City
Salt Lake City
Cheyenne
Des Moines
Lincoln
Chicago
Springfield
Columbus
Indianapolis
Annapolis
WASHINGTON D.C. 5.0 m
San Francisco
Denver
Topeka
St. Louis
Richmond
Charleston
Las Vegas
Los Angeles
Santa Fe
Oklahoma City
Little Rock
Nashville
Raleigh
Columbia
San Diego
Phoenix
El Paso
Dallas
Jackson
Montgomery
Atlanta
Savannah
Tallahassee
Jacksonville
Austin
Baton Rouge
Houston
New Orleans
Tampa
Orlando
Miami

San Antonio

NORTH ATLANTIC OCEAN

MEXICO
GULF OF MEXICO
BAHAMAS
CUBA

0 — 500 miles
0 — 500 km

Canada

Stretching from the Atlantic to the Pacific, Canada is the second-largest country in the world by landmass, comprising ten provinces and three territories. Large-scale European colonization began in the 17th century, with England and France the most active powers. In 1763, the French ceded control of their Canadian territory to Britain, though French influence remains strong, particularly in the province of Quebec. The country is officially bilingual, and over one-fifth of Canadians speak French as their first language. In 1867, the provinces confederated to form a dominion with domestic self-government. Canada became independent in 1931, although the British parliament could amend the Canadian Constitution until 1982. Due to migration from other continents, the country has become increasingly multicultural. A combination of abundant natural resources, such as oil and natural gas, a skilled labour force and free trade with the US (with whom it shares the world's longest two-country border) has helped make Canada one of the world's wealthiest and most developed countries.

QUEEN ELIZABETH ISLANDS

ARCTIC OCEAN

BEAUFORT SEA

ALASKA U.S.A.

GREENLAND SEA

BAFFIN BAY

VICTORIA ISLAND

BAFFIN ISLAND

Yukon R.

Great Bear Lake

Whitehorse

Echo Bay

Yellowknife

Baker Lake

Great Slave Lake

Rankin Inlet

Iqaluit

HUDSON STRAIT

LABRADOR SEA

PACIFIC OCEAN

COAST MTS

Prince Rupert

Dawson Creek

Churchill

HUDSON BAY

Prince George

Churchill R.

Thompson

Inukjuak

Edmonton

Nelson R.

Labrador City

NEWFOUNDLAND

Vancouver

Calgary

Saskatoon

St. John's

Victoria

Trail

Lethbridge

Lake Winnipeg

Charlottetown

Regina

Winnipeg

Moosonee

Fredericton

Amos

Quebec

Halifax

U.S.A.

Lake Superior

Moosonee

OTTAWA
1.3 m ★

Montreal

NOVA SCOTIA

Canada
45.42° N, 75.69° W
UTC –3.5 to UTC –8

U.S.A.

Michigan

Huron

Toronto
6.0 m

ATLANTIC OCEAN

☐ 9,984,670 km² (3,855,103 sq mi)	⚥ 80% European 20% others	🏦 Canadian Dollar (CAD)
○ Temperate in south, subarctic and arctic in north	☻ English, French	📈 $1.53 tn
👤 36,286,425	⛪ Christianity	📊 $42,158
⊞ 4/km² (10/sq mi)	⚖ Federal parliamentary system under constitutional monarchy	🚗 Automobile and aircraft manufacture, chemical manufacture, logging, oil and natural gas, agriculture and food processing
👥 1.2%		
☐ 82 : 18%	🌐 CON, G7, IMF, NAFTA, NATO, OAS, OECD, UN, WB, WTO	

Mexico

Arid to the north, tropical to the south and with a temperate central plateau, Mexico is made up of 31 states and a federal district around its capital, the most populous metropolitan area in the Western Hemisphere. It was home to several ancient civilizations including the Olmec, Toltec, Maya and Aztec. In 1521, Spanish conquistadors overthrew indigenous rule and established the Viceroyalty of New Spain, which extended into Central America and much of modern-day US territory. Mass migration from Europe led to the decimation of the indigenous population through disease. Declared in 1810, independence followed a war that lasted until 1821. A century of upheaval ensued, with political battles between liberals and conservatives, civil wars, uprisings, foreign interventions, two brief imperial periods and significant territorial losses. Despite ongoing violence with drug cartels and left-wing militants, the past century has been comparatively peaceful. Mexico's economy has advanced rapidly since the 1940s, industrializing and benefitting from offshore oil reserves discovered in 1976 and from NAFTA membership since 1994.

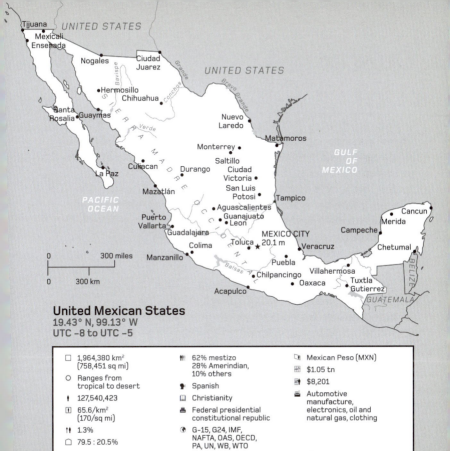

United Mexican States
19.43° N, 99.13° W
UTC -8 to UTC -5

☐ 1,964,380 km² (758,451 sq mi)	🏃 62% mestizo 28% Amerindian, 10% others	🖥 Mexican Peso (MXN)
O Ranges from tropical to desert	♟ Spanish	💵 $1.05 tn
✝ 127,540,423	📖 Christianity	💴 $8,201
🏙 65.6/km² (170/sq mi)	♨ Federal presidential constitutional republic	🚗 Automotive manufacture, electronics, oil and natural gas, clothing
🏃🏃 1.3%	⊕ G-15, G24, IMF, NAFTA, OAS, OECD, PA, UN, WB, WTO	
⌂ 79.5 : 20.5%		

Guatemala

The most populous country in Central America, Guatemala once lay at the heart of the Maya civilization, which established several extensive cities before going into decline in the tenth century, owing to a combination of climate change, disease and warfare. By the time the Spanish arrived in the early 16th century, the Maya had largely abandoned lowland cities and relocated to highland areas in the interior. Conquered by the Spanish in 1524, the country became part of the Mexican empire in 1821. Two years later it joined the separate Federal Republic of Central America along with Costa Rica, El Salvador, Honduras and Nicaragua. Full independence was achieved in 1839, after which the country vacillated between periods of conservative and liberal rule. In 1960, a 36-year civil war began, pitting the US-backed government against leftist rebels who tended to have indigenous support. By the time peace was made, in 1996, the conflict had claimed 200,000 lives, many of them civilian. The country is struggling to rebuild, and faces problems with high-level corruption, drug-smuggling activity and discrimination against Maya people.

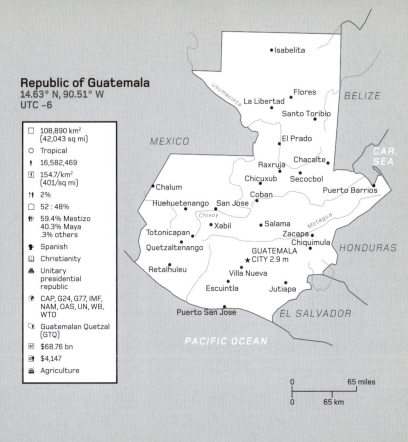

Republic of Guatemala
14.63° N, 90.51° W
UTC –6

- ☐ 108,890 km²
  (42,043 sq mi)
- ○ Tropical
- ⚲ 16,582,469
- ▣ 154.7/km²
  (401/sq mi)
- ⚧ 2%
- ◻ 52 : 48%
- ⚶ 59.4% Mestizo
  40.3% Maya
  .3% others
- ⚑ Spanish
- ▭ Christianity
- ⚏ Unitary
  presidential
  republic
- ◉ CAP, G24, G77, IMF,
  NAM, OAS, UN, WB,
  WTO
- ⬚ Guatemalan Quetzal
  (GTQ)
- ▨ $68.76 bn
- ▤ $4,147
- ▦ Agriculture

MEXICO

BELIZE

CAR.
SEA

HONDURAS

EL SALVADOR

PACIFIC OCEAN

•Isabelita

Usumacinta

•Flores

La Libertad•

Santo Toribio•

El Prado•

Raxruja•

Chicuxub•   Secocbol•   Chacalte•

Puerto Barrios•

Chalum•

Coban•

Huehuetenango•   San Jose•

Chixoy

Salama•

Motagua

Xabil•   Zacapa•

Totonicapan•   Chiquimula•

Quetzaltenango•   ★ GUATEMALA
CITY 2.9 m

Retalhuleu•   Villa Nueva•

Escuintla•   Jutiapa•

Puerto San Jose•

| 0 | | 65 miles |
| 0 | | 65 km |

El Salvador

The smallest nation in the mainland Americas, as well as the most densely populated, El Salvador is largely characterized by mountainous terrain with a plain fringing its Pacific coast. Spain began its conquest of the country in 1524, but faced a long struggle with the indigenous population, particularly from the Pipil and the Lenca peoples. After finally establishing dominance in 1540, Spain continued to rule until 1821. Two years later El Salvador became part of the Federal Republic of Central America, and when that body dissolved in 1841, the country gained full independence, becoming increasingly reliant on coffee exports. A tumultuous 20th century saw a succession of military regimes (1931–79) followed by a 12-year civil war between the government and left-wing guerrillas that led to the deaths of 75,000 people. Democratic rule has since been restored and the government has signed several free-trade agreements to encourage economic growth and diversification from reliance on coffee. However, the country still suffers from periodic natural disasters, organized crime and gang-related violence.

Republic of El Salvador
13.69° N, 89.22° W
UTC –6

☐ 21,040 km² (8,124 sq mi)	♟ 86.3% mestizo 12.7% white 0.9% others	⊕ CAP, G77, IMF, OAS, UN, WB, WTO
O Tropical	♟ Spanish	⌨ United States Dollar (USD)
♦ 6,344,722	☐ Christianity	▦ $26.80 bn
⊞ 306.2/km² (793/sq mi)	♟ Unitary presidential constitutional republic	▦ $4,224
♦♦ 0.5%		⛏ Agriculture, clothing
◻ 67.2 : 32.8%		

Belize

Belize has a diverse landscape that includes mangrove swamps, lagoons and rainforest; its coast is covered by a section of the Mesoamerican Reef, the largest such body in the Northern Hemisphere. Belize was the site of several Maya city-states, and although these went into decline from 900 CE, the local population resisted Spanish incursions during the 16th century. The first European colonists were British pirates and loggers, who settled coastal areas from the mid-17th century, fighting off Spanish attempts to dislodge them that continued until 1798. The area became the colony of British Honduras in 1862, and changed its name to Belize in 1973. Disputes over Guatemala's claim to the territory delayed independence until 1981; the issue remains unresolved, and Britain maintains a military presence in the country, partly to protect Belizean sovereignty. As an independent nation, Belize has become a stable parliamentary democracy, although poverty and drug-related crime remain problematic. The economy is based on agricultural exports and tourism, both of which are vulnerable to global market conditions and hurricanes.

Belize
17.25° N, 88.76° W
UTC –6

☐	22,970 km² (8,869 sq mi)
○	Tropical
♦	366,954
▣	16.1/km² (41.7/sq mi)
♇	2.1%
◖	43.8 : 56.2%
♚	52% mestizo 23% Creole 10% Maya 15% others
♠	English
◻	Christianity
♣	Unitary parliamentary under constitutional monarchy
♁	CARICOM, CON, G77, IMF, NAM, OAS, UN, WB, WTO
◻	Belize Dollar (BZD)
▨	$1.77 bn
▩	$4,811
☶	Agriculture, tourism

MEXICO

Pucte

Orange Walk

Rancho

AMBERGRIS CAYE

Water Bank

Belize City
62 k

GUATEMALA

Belize

Rockville

TURNEFFE ISLANDS

★ BELMOPAN
17 k

CARIBBEAN SEA

Dangriga

Locust Bank

Monkey River Town

San Antonio

Punta Gorda

GULF OF HONDURAS

GUATEMALA

0   65 miles

0   65 km

Honduras

Honduras boasts coastlines on both the Caribbean and Pacific, yet the majority of its population inhabits the rugged highlands of the interior. Home to several Mesoamerican peoples, including the Maya, Honduras was conquered by Spain in the early 16th century. Spanish rule ended in 1821; initially, Honduras became part of the Mexican empire before splitting off to join the Federal Republic of Central America in 1823. Complete independence was achieved in 1838, after which the country was highly unstable and saw a rapid turnover of presidents and numerous internal revolts. During the 20th century, the US became increasingly influential, initially to protect the interests of American fruit companies there and, in the 1980s, to use the country as a base for the right-wing guerrillas they supported in Central America. Although Honduras is nominally under civilian rule, the military remains a strong political influence and even removed the elected president during a coup in 2009. The country endures high levels of debt and poverty, as well as vulnerability to hurricanes and some of the highest murder rates in the world.

Republic of Honduras
14.07° N, 87.19° W
UTC −6

☐ 112,490 km² (43,433 sq mi)	⚧ 0.1%	⚙ CAP, G77, IMF, NAM, OAS, UN, WB, WTO
○ Subtropical in lowlands, temperate in mountains continental	⌂ 55.3 : 44.7%	🗎 Honduran Lempira (HNL)
⚥ 9,112,867	⚥ 90% mestizo, 10% others	▦ $ 21.52 bn
⊡ 81.4/km² (211/sq mi)	☗ Spanish	▤ $2,361
	☐ Christianity	▰ Agriculture
	⚖ Presidential republic	

Nicaragua

The Spanish named Nicaragua after an indigenous chief they called Nicarao. The country is divided between valleys and lakes in the west and lowlands and lagoons in the east, separated by mountainous terrain in the centre. The Spanish conquered most of the country in 1523–24, but could not subdue the Miskito people who lived in the eastern region known as the Mosquito Coast; this area became a British protectorate during the mid-17th century. Spanish rule ended in 1821 and Nicaragua was part of the Mexican empire and then the Federal Republic of Central America, prior to achieving full independence in 1838. Britain ceded the Mosquito Coast in 1860. Nicaragua has endured political instability, military dictatorship and foreign interference – most recently when the left-wing Sandinista movement won power in 1979 and were opposed by right-wing Contra rebels funded by the US. This led to warfare that lasted until 1990. Since then Nicaragua has largely been peaceful and democratic, though it has suffered from several serious hurricanes and earthquakes, as well as high levels of unemployment and poverty.

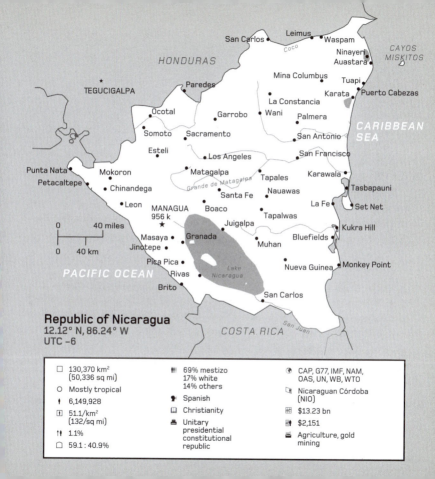

Republic of Nicaragua
12.12° N, 86.24° W
UTC –6

☐	130,370 km² (50,336 sq mi)	�	69% mestizo 17% white 14% others	☻	CAP, G77, IMF, NAM, OAS, UN, WB, WTO
O	Mostly tropical	☻	Spanish	⌨	Nicaraguan Córdoba (NIO)
☗	6,149,928	☖	Christianity	📈	$13.23 bn
⊞	51.1/km² (132/sq mi)	♟	Unitary presidential constitutional republic		$2,151
↕	1.1%			⚒	Agriculture, gold mining
☐	59.1 : 40.9%				

Costa Rica

Covering just 0.03 per cent of Earth's landmass, but home to 5 per cent of its biodiversity, Costa Rica lives up to the name European explorers gave it in 1502, which means 'rich coast'. The Spanish settled here in the 1560s, but the area had no mineral resources, so remained mostly agricultural and sparsely populated. Spain relinquished control in 1821, and Costa Rica was part of the Mexican empire before joining the Federal Republic of Central America in 1823. After full independence in 1838, the country became a major coffee exporter, thanks to its fertile soil. Although Costa Rica has generally been peaceful and stable, it suffered a civil war in 1948, when the government annulled elections that brought a socialist candidate to office. Opposition forces were victorious; the new regime introduced sweeping reforms including the abolition of the armed forces. The country has become increasingly prosperous by diversifying its economy and encouraging foreign investment through free-trade zones. It is one of the greenest countries in the world; by 2016, almost all of its energy was generated by renewable sources.

Republic of Costa Rica
9.93° N, 84.09° W
UTC −6

☐ 51,100 km² (19,730 sq mi)	⚤ 84% white/mestizo 16% others	✪ G77, IMF, OAS, UN, WB, WTO
○ Tropical to subtropical	✸ Spanish	⬛ Costa Rican Colón (CRC)
⚥ 4,857,274	⌂ Christianity	▦ $57.43 bn
▦ 95.1/km² (246/sq mi)	⚒ Unitary presidential constitutional republic	▦ $11,825
⚤ 1%		⚒ Agriculture, tourism, medical devices
◯ 77.7 : 22.3%		

Cuba

The largest nation in the Caribbean by both area and population, Cuba has a landscape of which two-thirds is covered by plains; there are also areas of swamp and highlands. Spain claimed Cuba in 1492, establishing settlements in 1511; the indigenous population was virtually eradicated through violence, disease and displacement. The Spanish brought in African slaves to work on plantations – primarily sugar – and slavery was not abolished until 1886. By the 19th century, dissatisfaction with the colonial regime led to several revolts culminating in independence from Spain in 1898. The United States supported the rebels and placed Cuba under American administration; independence followed in 1902. In the 1950s, Fidel Castro led a communist uprising against the incumbent military dictatorship, winning power in 1959 and ruling until 2008. His nationalization of US businesses and close ties to the USSR led to numerous American attempts to overthrow him and the imposition of long-term trade sanctions. Cuba remains a one-party socialist state although, since 2011, the economy has been liberalized and political prisoners have been released.

U.S.

GULF OF MEXICO

STRAITS OF FLORIDA

ATLANTIC OCEAN

THE BAHAMAS

GREAT BAHAMA BANK

HAVANA
2.1 m ★

San Cristobal
Pinar del Rio

Matanzas

Colón

Santa Clara

Cienfuegos

Morón

Sancti Spiritus

Maria la Gorda

ISLA DE LA JUVENTUD

Camagüey

Tunas

Holguin

Moa

Baracoa

Manzanillo

Santiago de Cuba

Guantánamo

0   100 miles

0   100 km

CAYMAN ISLANDS (U.K.)

U.S. NAVAL BASE GUANTÁNAMO BAY

CARIBBEAN SEA

JAMAICA

HAITI

Republic of Cuba
23.11° N, 82.37° W
UTC −5

☐ 109,880 km² (42,425 sq mi)	⚘ Spanish	Cuban Peso (CUP) and Cuban Convertible Peso (CUC)
○ Tropical	⌂ Christianity	$87.13 bn (2015)
✝ 11,475,982	⚒ Unitary one-party socialist republic	$7,602 (2015)
⊡ 110.3/km² (286/sq mi)	⊕ G77, NAM, OAS (suspended 1962–2009, has chosen not to return), UN, WTO	Agriculture, tourism
↟↡ 0.1%		
◠ 77.2 : 22.8%		
⚲ 64% white 27% mixed 9% black		

Panama

Bridging North and South America, the Isthmus of Panama is a narrow strip of land that separates the Caribbean and the Pacific. During the early 16th century, Spain colonized the area, subjugating its indigenous population and ruling until 1821. Panama then joined Gran Colombia, at the time made up of former Spanish colonies in northern South America. This federation collapsed in 1830 and Panama then became part of Colombia, until seceding in 1903. That same year Panama signed a treaty with the US giving the latter the right to build and control a canal across the country. The 80-km (50-mile) canal was completed in 1914, and became one of the most important waterways in the world. Following an American invasion that ended Manuel Noriega's military dictatorship in 1989, Panama transitioned to democracy and became increasingly prosperous. The country was given control of the canal in 1999; from 2007–16 it widened the canal, doubling its capacity. Panama's economy is dominated by shipping and transit services, although the country is also a major centre of offshore banking.

Republic of Panama
8.98° N, 79.52° W
UTC –5

☐ 75,420 km² (29,120 sq mi)	✤ Spanish	📖 $55.19 bn
○ Tropical maritime	📖 Christianity	📄 $13,680
♦ 4,034,119	⚒ Unitary presidential constitutional republic	🏭 Transport and logistics services, banking
🗓 54.3/km² (141/sq mi)		
‼ 1.6%	🌐 CAP, G77, IMF, NAM, OAS, UN, WB, WTO	
⬠ 66.9 : 33.1%		
♛ 65% mestizo 12% indigenous 23% others	💱 Panamanian Balboa (PAB), United States Dollar	

The Bahamas

Located on a coral archipelago, The Bahamas is made up of over 700 islands; around 30 of them are inhabited, with two-thirds of the population living on the largest, New Providence. San Salvador Island, to the southeast, was probably the site at which Christopher Columbus first landed in the New World, in 1492. During the three decades that followed, the indigenous Lucayan – a branch of the Taíno ethnic group – were eradicated through a combination of Western diseases and transportation to work in mines on Hispaniola. Otherwise, European powers showed little interest in The Bahamas until 1648, when English settlers migrated there. From 1670, the islands were placed under the control of the Province of Carolina (in modern-day United States), and it became an infamous haunt of pirates. As a result, Britain extended colonial government over The Bahamas in 1717, ruling until independence was granted in 1973. Despite its thin and infertile soil, the country has benefitted from its close proximity to the US and has become the wealthiest nation in the Caribbean, largely thanks to tourism and offshore banking.

☐	13,880 km² (5,359 sq mi)	📖	Christianity
○	Tropical marine	⚖	Unitary parliamentary under constitutional monarchy
👤	391,232		
👥	39.1/km² (101/sq mi)	🏛	CARICOM, CON, G77, IMF, NAM, OAS, UN, WB, WTO (observer)
👫	1.1%		
☐	83 : 17%	💴	Bahamian Dollar (BSD)
👪	91% black 9% others	💵	$9.05 bn
👤	English Bahamian Creole	💰	$23,124
		🚂	Tourism, banking

Cornishtown

Hope Town

Freeport

GRAND BAHAMA

ABACO

STRAITS OF FLORIDA

Dunmore Town

NASSAU 267 k

ELEUTHERA

Nicholls Town

Governor's Harbor

NEW PROVIDENCE

Andros Town

Arthur's Town

New Bight

Cockburn Town

CAT ISLAND

SAN SALVADOR

Congo Town

ANDROS ISLAND

Commonwealth of the Bahamas
25.05° N, 77.36° W
UTC −5

LONG ISLAND

Georgetown

GREAT EXUMA

Clarence Town

GREAT BAHAMA BANK

Colonel Hill

ATLANTIC OCEAN

CROOKED ISLAND

MAYAGUANA

RAGGED ISLAND

Abraham's Bay

ACKLINS ISLAND

0 — 125 miles

0 — 125 km

GREAT INAGUA

Matthew Town

Jamaica

The original inhabitants of this Caribbean island – the Taíno people – were wiped out through enslavement and disease, in the two centuries after Spain laid claim to Jamaica in 1494. Britain captured the island in 1655 and it became a centre of the slave trade and a major sugar producer. However, thousands of slaves, known as Maroons (a derivation of the Spanish *cimarrón*, meaning 'untamed') escaped into the mountainous interior, and a series of conflicts resulted in the colonial authorities recognizing their autonomy in 1740. Jamaica joined the West Indies Federation in 1958, but after that organization collapsed in 1962, Jamaica won full independence – the first British Caribbean colony to do so. Initially, the economy boomed, thanks to bauxite mining and tourism, but during the 1970s and 1980s the country suffered recession and political violence. Despite a comparatively large and diversified economy in regional terms, growth in recent decades has been slow, with gang violence, corruption and poverty remaining troublesome issues. Nevertheless, Jamaica's musical and sporting traditions give it a global cultural influence that far outweighs its size.

CARIBBEAN SEA

Montego Bay
Falmouth
Lucea
Reading
Runaway Bay
Saint Anns Bay
Oracabessa
Montpelier
Ocho Rios
Port Maria
Negril
Moneague
Annotto Bay
Savanna la Mar
Port Antonio
Christiana
Frankfield
Linstead
Chapelton
Bog Walk
Black River
Half Way Tree
Mandeville
Spanish Town
May Pen
Old Harbour
Alligator Pond
KINGSTON
588 k
Morant Bay

CARIBBEAN SEA

| 0 | | 24 miles |
| 0 | 24 km | |

Jamaica
18.02° N, 76.81° W
UTC −5

☐ 10,990 km² (4,243 sq mi)	⚕ 92% black 6% mixed 2% others	⊛ CARICOM, CON, G-15, G77, IMF, NAM, OAS, UN, WB, WTO	
○ Tropical	☙ English	⬡ Jamaican Dollar (JMD)	
✝ 2,881,355	⬚ Christianity	▦ $14.03 bn	
⊞ 266.1/km² (689/sq mi)	⬥ Unitary parliamentary system under constitutional monarchy	▥ $4,868	
⬗ 0.3%		⬛ Tourism, bauxite/ alumina mining	
☐ 55 : 45%			

Haiti

Occupying the western side of the Caribbean island of Hispaniola, Haiti is a mountainous country with river valleys and coastal plains. The Spanish established colonial rule on the island in 1492, virtually wiping out the indigenous Taíno population through enslavement and disease. The French gained control of the western third of Hispaniola in 1697, naming their new colony Saint-Domingue and establishing a slave-based plantation system. A slave rebellion erupted in 1791 and, after a protracted struggle, French rule was cast off in 1804, and the Republic of Haiti declared. France only recognized Haitian independence in 1825, imposing reparations that were not paid off until 1947. Foreign involvement continued when the United States occupied the country to protect its economic interests from 1915–34. Haiti has veered between periods of instability and authoritarianism, and did not hold free and fair elections until 1990. It remains the poorest and least-developed nation in North America; it is still recovering from the 2010 earthquake that struck Port-au-Prince, killing 300,000 and causing severe damage to the economy.

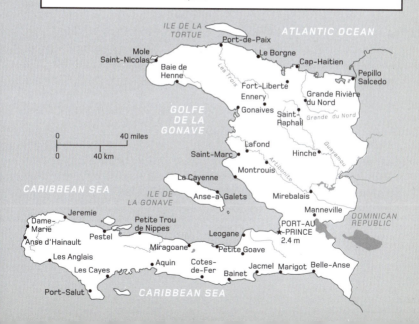

Republic of Haiti
18.59° N, 72.31° W
UTC −5

☐ 27,750 km² (10,714 sq mi)	☐ 59.8 : 40.2%	⏣ CARICOM, G77, IMF, NAM, OAS, UN, WB, WTO
○ Mostly tropical	⋔ 95% black 5% others	💵 Haitian Gourde (HTG)
⊹ 10,847,334	☻ French Haitian Creole	▨ $8.02 bn
⊡ 393.6/km² (1,019/sq mi)	▯ Christianity	▨ $740
⊹⊹ 1.3%	⛁ Unitary semi-presidential republic	⚒ Agriculture

ILE DE LA TORTUE

ATLANTIC OCEAN

Port-de-Paix

Mole Saint-Nicolas

Le Borgne

Cap-Haitien

Baie de Henne

Pepillo Salcedo

Les Trois

Fort-Liberte

Ennery

Grande Rivière du Nord

GOLFE DE LA GONAVE

Gonaives

Saint-Raphali

Grande du Nord

Lafond

Saint-Marc

Hinche

Guayamou

Montrouis

Artibonite

La Cayenne

CARIBBEAN SEA

ILE DE LA GONAVE

Anse-a-Galets

Mirebalais

Manneville

DOMINICAN REPUBLIC

Jeremie

Petite Trou de Nippes

Dame-Marie

Pestel

Leogane

PORT-AU-PRINCE
2.4 m

Anse d'Hainault

Miragoane

Petite Goave

Les Anglais

Aquin

Cotes-de-Fer

Jacmel

Marigot

Belle-Anse

Les Cayes

Bainet

Port-Salut

CARIBBEAN SEA

0 — 40 miles
0 — 40 km

Dominican Republic

The eastern side of the island of Hispaniola, the Dominican Republic has fertile valleys and four major mountain ranges. It was the site of Santo Domingo, the first permanent Western colony in the Americas, founded by Spain in 1496. Spanish control of Hispaniola was threatened by other European powers during the 16th and 17th centuries; in 1697, Spain granted France the western part of the island, which went on to become Haiti. Santo Domingo won independence in 1821, but was annexed by Haiti the following year. Dominican sovereignty was restored in 1844 and, aside from a return to Spanish rule from 1861-65, it has been an independent republic since that time. The United States became increasingly influential, occupying the Dominican Republic from 1916–24 following internal disorder. Rafael Trujillo ruled from 1930–61, his dictatorial regime overseeing the deaths of 50,000 people. Since the 1960s, the Dominican Republic has returned to democracy; it still suffers from corruption and natural disasters, and continues to rely heavily on the US as a trading partner and for tourism and remittances.

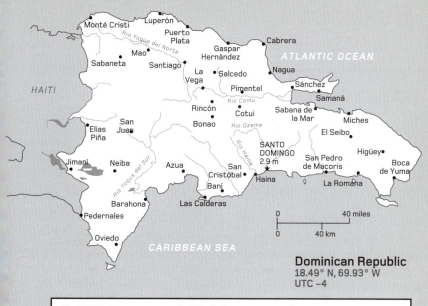

Monté Cristi
Luperón
Puerto Plata
Cabrera
ATLANTIC OCEAN
Gaspar Hernández
Mao
Sabaneta
Santiago
La Vega
Salcedo
Nagua
Sánchez
Pimentel
Samaná
Rio Yaque del Norte
Rio Comu
Rincón
Cotui
Sabana de la Mar
Miches
Bonao
Rio Ozama
El Seibo
HAITI
Elías Piña
San Juan
Higüey
Boca de Yuma
Jimaní
Neiba
Azua
Rio Haina
SANTO DOMINGO 2.9 m
San Pedro de Macoris
San Cristóbal
Haina
Baní
La Romana
Rio Yaque del Sur
Barahona
Las Calderas
Pedernales
Oviedo
CARIBBEAN SEA

Dominican Republic
18.49° N, 69.93° W
UTC −4

☐ 48,670 km² (18,792 sq mi)	♛ 73% mixed 16% white 11% black	🗠 Dominican Peso (DOP)
O Tropical maritime	♟ Spanish	🖾 $71.58 bn
♙ 10,648,791	🕮 Christianity	🖹 $6,722
⊞ 220.4/km² (571/sq mi)	🜲 Unitary presidential republic	🖴 Agriculture, tourism, mineral mining (particularly ferronickel and gold)
♙♙ 1.1%	◉ CAP, G77, IMF, NAM, OAS, UN, WB, WTO	
⌂ 79.8 : 20.2%		

St Kitts
and Nevis

With the smallest area and population in the Caribbean, St Kitts and Nevis is made up of two mountainous islands separated by a 3-kilometre (1.9-mile) channel. European colonists did not arrive until the early 17th century; in 1623, St Kitts was partitioned between the English and French, while the English alone settled Nevis, in 1628. The Europeans killed off the local Carib population, imported African slave labour and established sugar plantations. The British fought off the Dutch and Spanish for control of Nevis, and in 1783 won sole control of St Kitts. Along with Anguilla, the islands were administratively joined into a single colony in 1882, which became a province of the West Indies Federation from 1958–62. St Kitts and Nevis became the last British Caribbean colony to achieve independence, in 1983 (Anguilla split off in 1967 and remains under the sovereignty of the United Kingdom). In 2005, sugar production ceased in St Kitts and Nevis after decades of losses; by that time its economy had largely shifted towards tourism, banking and offshore financial services.

Saint Paul's
Sadlers
Mension
SAINT KITTS
Sandy Point Town
Cayon
ATLANTIC OCEAN
Half Way Tree
Verchild's
Old Road Town
Monkey Hill Village
BASSETERRE
14 k
Challengers
Boyd's
CARIBBEAN SEA

0   4 miles
0   4 km

**Federation of
Saint Kitts and Nevis**
17.30° N, 62.72° W
UTC –4

☐ 260 km² (100 sq mi)	🏛 Federal parliamentary system under constitutional monarchy
○ Tropical	
✝ 54,821	
⊞ 210.9/km² (546/sq mi)	⊛ CARICOM, CON, G77, IMF, NAM, OAS, UN, WB, WTO
↟↟ 1.0%	
◠ 32.2 : 67.8%	💴 East Caribbean Dollar (XCD)
⋔⋔ 93% black 7% others	💹 $916.89 m
🕭 English	🏦 $16,725
▨ Christianity	🚢 Tourism, banking

Newcastle
Burnaby
Whitehall
Cotton Ground
NEVIS
Charlestown
New River
Zetlands
Brown Hill

Trinidad and Tobago

U nlike other Caribbean islands, Trinidad was once part of the South American landmass, before being separated by rising sea levels 10,000 years ago. Originally inhabited by Carib and Arawak people, it was claimed by Spain in 1498, then captured by Britain in 1797. Following the abolition of slavery in 1838, plantation owners turned to indentured labour, particularly from the Indian subcontinent. Tobago, 30 km (19 miles) northeast of Trinidad, was settled by the Dutch in the 1630s. It changed hands between European powers several times before being ceded to Britain in 1814. Hurricanes contributed to the collapse of Tobago's sugar industry, prompting economic difficulties that led to unification as a single colony with Trinidad, in 1889. Together they joined the short-lived West Indies Federation in 1958, and won independence in 1962. By that time oil (discovered in 1857) had become the country's staple industry. Trinidad and Tobago are now the Caribbean's leading producers of natural gas and oil, although they have sought to lessen their reliance on their export by branching out into other fields, such as chemicals and metals.

- ☐ 5,130 km² (1,981 sq mi)
- ○ Tropical
- ♀ 1,364,962
- ⊞ 266.1/km² (690/sq mi)
- ♀♀ 0.4%
- ◖ 8.4 : 91.6%
- ♔♔ 35% East Indian 34% African 23% Mixed 8% others
- ♟ English

- ◫ Christianity
- ♜ Unitary parliamentary under constitutional republic
- ⊕ CARICOM, CON, G24, G77, IMF, NAM, OAS, UN, WB, WTO
- ⬚ Trinidad and Tobago Dollar (TTD)
- ⊞ $20.99 bn
- ⊞ $15,377
- ⬛ Petroleum and natural gas

Charlotteville

Moriah

TOBAGO

Plymouth

Roxborough

Scarborough

Canaan

| 0 | | 12 miles |
| 0 | 12 km | |

Toco

Maracas

Maraval

Saint Joseph

PORT-OF-SPAIN 37 k

Arouca

Arima

Guaico

Sangre Grande

Chaguanas 84 k

TRINIDAD

Republic of Trinidad and Tobago
10.65° N, 61.50° W
UTC –4

Tabaquite

CARIBBEAN SEA

San Fernando

Rio Claro

Brighton

La Brea

Princes Town

Tableland

Pierreville

Point Fortin

Debe

ATLANTIC OCEAN

Siparia

Basse Terre

Guayaguayare

Fullarton

San Francique

Moruga

Antigua and Barbuda

Made up of two main islands just under 40 kilometres (25 miles) apart, Antigua and Barbuda were sighted by Christopher Columbus in 1493. The English began to settle Antigua from 1632. Their colonization of Barbuda began in 1678, and seven years later it was leased to the Codrington family, who controlled it until 1860. Sugar plantations using slave labour formed the foundation of the economy on both islands, even after emancipation in 1834. The country moved towards domestic sovereignty from the mid-20th century; from being a member of the West Indies Federation from 1958–62, it gained internal self-governance in 1967 and full independence in 1981. The economy has since moved away from sugar production, which had ended by 1972 owing to foreign competition. In recent years, American trading restrictions have led to a decline in the country's online gambling industry and tourism has become vital as a result. Tropical storms pose a significant threat; in 2017, Hurricane Irma struck Barbuda, destroying 95 per cent of residences and requiring its population to evacuate to Antigua.

Antigua and Barbuda
17.13° N, 61.85° W
UTC −4

▢	440 km² (170 sq mi)
○	Tropical maritime
✦	100,963
⊞	229.5/km² (594/sq mi)
↟↟	1.0%
◠	23.4 : 76.6%
⩗	87% black 13% others
⚥	English
▣	Christianity
⚒	Parliamentary democracy under constitutional monarchy
⊛	CARICOM, CON, G77, IMF, NAM, OAS, UN, WB, WTO
▱	East Caribbean Dollar (XCD)
⊞	$1.45 bn
⊟	$14,353
⊟	Tourism

Codrington

BARBUDA

ATLANTIC OCEAN

0   8 Miles
0   8 KM

CARIBBEAN SEA

Cedar Grove

SAINT JOHN'S
22 K

Parham

ANTIGUA

Willikies Village

All Saints

Bolands

Freetown

Sweets

Carlisle

English Harbour Town

Grenada

The most southerly of the Windward Islands, Grenada has a mountainous, forested interior and rich volcanic soil. The country also includes the dependency of Carriacou and Petite Martinique, two islands in the southern part of the Grenadines archipelago. During the mid-17th century, French colonists defeated the indigenous Caribs, and established sugar and indigo plantations. Britain gained control of the island in 1763 and, aside from a brief reintroduction of French rule from 1779–83, their colonial regime continued for over two centuries. Nutmeg, introduced in 1843, became Grenada's economic backbone (and is represented on the national flag); the country grows over one-third of the world's supply of the spice. After joining the West Indies Federation from 1958–62, Grenada won independence in 1974. Five years later the elected leader was deposed in a coup that established a single-party socialist state closely allied with Cuba. In 1983, an American-led invasion force overthrew this regime, restoring the previous democratic political system; since then the country has been politically stable.

☐	340 km² (131 sq mi)
○	Tropical
✝	107,317
⊞	315.6/km² (817/sq mi)
↑↑	0.5%
◠	35.6 : 64.4%
ⅰⅰ	82% African 13% mixed 5% others
☻	English, Creole French
▭	Christianity
♣	Parliamentary democracy under constitutional monarchy
⊕	CON, G77, IMF, NAM, OAS, UN, WB, WTO
◔	East Caribbean Dollar (XCD)
▩	$1.02 bn
≋	$9,469
🍴	Tourism, agriculture

Grenada
12.06° N, 61.75° W
UTC −4

PETIT MARTINIQUE

Hillsborough
Grand Bay

CARRIACOU

CARIBBEAN SEA

RONDE ISLAND

Sauteurs

Victoria
Tivoli
Gouyave

Grenville

Grand Roy

Marquis

ST.GEORGE'S
★ 38 k
Belmont
Calivigny
Saint Davids

ATLANTIC OCEAN

0		4 miles
0	4 km	

Dominica

A volcanic island with a central mountain ridge, Dominica has landscape of rainforest and woodland that is home to numerous rare species, including the sisserou parrot, which appears on the national flag. It was the last Caribbean island to be colonized – by the French in 1727 – owing to resistance from the local Carib population. The French established plantations that used slave labour during the 18th century but, by 1805, Britain had fought off France to win control of the island. In 1838, Dominica became the first and only British Caribbean colony to elect a majority of black people to its legislature, but power soon became re-entrenched in the white landowning elite. By the mid-20th century, a democratic system had been put in place, before the country's brief membership of the West Indies Federation from 1958–62. Dominica won independence in 1978 and, in 1981, survived an attempted coup led by foreign mercenaries. The main threat to the country comes from tropical storms; in 2017, Hurricane Maria destroyed or damaged 90 per cent of the structures on the island.

Commonwealth of Dominica
15.31° N, 61.38° W
UTC −4

- ☐ 750 km²
  (290 sq mi)
- ○ Tropical
- 🕯 73,543
- ⊞ 98.1/km²
  (254/sq mi)
- 🕯🕯 0.5%
- ◠ 69.8 : 30.2%
- 🕯🕯 87% black
  13% others
- 🕯 English
- ▭ Christianity
- ♜ Unitary
  parliamentary
  republic
- ⊕ CARICOM, CON, G77,
  IMF, NAM, OAS, UN,
  WB, WTO
- ▢ East Caribbean
  Dollar (XCD)
- ▦ $525.42 m
- ▦ $7,144
- ▦ Agriculture, tourism

ATLANTIC OCEAN

Portsmouth

Marigot

Salisbury

CARIBBEAN SEA

Pont Casse

Rosalie

Laudat

ROSEAU
15 k

Pointe Michal

Berekua

| 0 | | 6 miles |
| 0 | | 6 km |

St Vincent and the Grenadines

The mountainous main island of St Vincent makes up more than 90 per cent of this country's landmass, with the remaining territory consisting of the Grenadines, a chain of islands and cays. During the 17th and 18th centuries, European powers battled each other for control of the country, although the indigenous Carib thwarted their attempts to establish settlements until 1719. The British gained control in 1783, and after a series of rebellions in the 1790s, the Carib population was deported. Until abolition in 1834, African slaves provided labour for the colony's plantations; thereafter, they were supplemented by indentured labourers from Portugal and the East Indies. Natural disaster struck St Vincent in 1812 and 1902 when the volcano La Soufrière erupted, causing thousands of deaths and serious damage. After a short spell in the Federation of the West Indies (1958–62), domestic rule was won in 1969 and independence in 1979. Today, the nation suffers from a small economic base and has attempted to expand from a reliance on bananas towards tourism and offshore financial services.

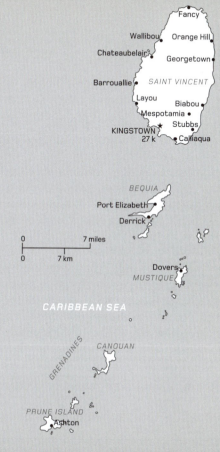

Saint Vincent and the Grenadines

13.16° N, 61.22° W
UTC −4

☐	390 km² (151 sq mi)
○	Tropical
✝	109,643
▦	281.1/km² (728/sq mi)
⚥	50.9 : 49.1%
◻	0.2%
👪	66% black 19% mixed 15% others
🗣	English
▢	Christianity
⚒	Parliamentary democracy under constitutional monarchy
◉	CARICOM, CON, G77, IMF, NAM, OAS, UN, WB, WTO
▭	East Caribbean Dollar (XCD)
▦	$770.80 bn
▧	$7,030
▰	Agriculture, tourism

Fancy
Wallibou · Orange Hill
Chateaubelair · · Georgetown
Barrouallie · SAINT VINCENT
Layou · · Biabou
Mespotamia · Stubbs
KINGSTOWN · · Calliaqua
27 k

BEQUIA
Port Elizabeth ·
Derrick ·

0	7 miles
0	7 km

Dovers
MUSTIQUE

CARIBBEAN SEA

GRENADINES · CANOUAN

PRUNE ISLAND
Ashton

St Lucia

The volcanic island of St Lucia has an interior featuring mountains and rainforest and a coast with numerous natural harbours. Carib peoples arrived here around 800 CE, forcing out earlier Arawak settlers. Europeans followed in the early 17th century, but attempts at colonization were frustrated by disease and attacks from the Carib. France finally established a colony in 1650, introducing sugar plantations and slavery. During the late 18th century, France and Britain battled for control of the island, which was ceded to the British in 1814. With slavery abolished from 1834, the colony slowly moved towards democracy, introducing an elected representative body in 1924, and universal adult suffrage in 1951. St Lucia joined the West Indies Federation in 1958; following its collapsed in 1962, the country gained domestic self-government in 1967 and full independence in 1979. Agriculture continues to be important, although tropical fruit, particularly bananas, has replaced sugar. This sector is highly vulnerable to the hurricanes that frequently strike the island, making its offshore banking and tourism sectors even more essential.

Saint Lucia
14.01° N, 60.99° W
UTC –4

- ☐ 620 km² (239 sq mi)
- ○ Tropical
- ♦ 178,015
- ⊞ 291.8/km² (756/sq mi)
- ♯♯ 0.5%
- ◯ 18.5 : 81.5%
- ♔♔ 85% black
  11% mixed
  4% others
- ♥ English
- ☐ Christianity
- ☰ Parliamentary democracy under constitutional monarchy
- ◉ CARICOM, CON, G77, IMF, NAM, OAS, UN, WB, WTO
- ☐ East Caribbean Dollar (XCD)
- ▦ $1.38 bn
- ▦ $7,744
- ☰ Tourism, agriculture

CARIBBEAN SEA

Gros Islet

Dauphin

CASTRIES
22 k

Grand Anse

Sans Soucis

La Croix Maingot

Anse La Raye

Canaries

Dennery

Praslin

Mon Repos

Soufriere

Micoud

Desruisseau

Londonderry

Choiseul

Laborie

Vieux Fort

ATLANTIC OCEAN

0 — 4 miles

0 — 4 km

Barbados

The easternmost Caribbean nation, Barbados has the highest population density in all the Americas. Amerindian, Arawak and Carib peoples settled here from the 6th to the 13th centuries, but in the early 16th century the island became depopulated as Spanish slaving raids led to the capture or flight of its inhabitants. Since Barbados was comparatively remote and had no precious metals, it remained deserted until claimed by England in 1625. The introduction of sugar in the mid-17th century transformed the island's fortunes; the English brought in African slaves to work the plantations and, by the end of the 18th century, there were 80,000 on the island. Even after the abolition of slavery in 1834, the white elite remained dominant until the 1940s and 1950s, when electoral reform finally opened up the vote to the majority of the population. After four years as part of the West Indies Federation, a union of British Caribbean colonies, Barbados won full independence in 1962. Since then, it has broadened its economy into tourism and, since 1985, offshore financial services.

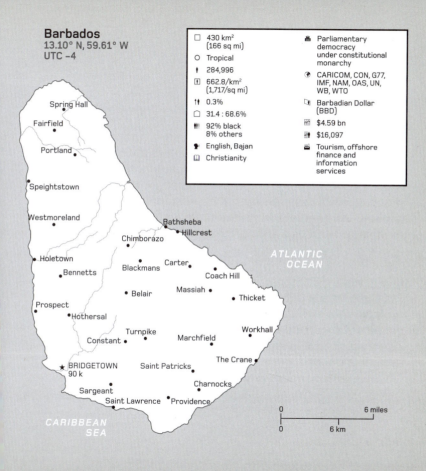

Barbados
13.10° N, 59.61° W
UTC −4

☐ 430 km²
(166 sq mi)

○ Tropical

✝ 284,996

⊞ 662.8/km²
(1,717/sq mi)

⇅ 0.3%

⌂ 31.4 : 68.6%

⋔ 92% black
8% others

♟ English, Bajan

📖 Christianity

♜ Parliamentary
democracy
under constitutional
monarchy

◉ CARICOM, CON, G77,
IMF, NAM, OAS, UN,
WB, WTO

💱 Barbadian Dollar
(BBD)

▦ $4.59 bn

▨ $16,097

🚢 Tourism, offshore
finance and
information
services

Spring Hall

Fairfield

Portland

Speightstown

Westmoreland

Bathsheba
Hillcrest

Chimborazo

Holetown

Bennetts

Carter

Blackmans

Coach Hill

Belair

Massiah

Thicket

Prospect

Hothersal

Turnpike

Workhall

Constant

Marchfield

★ BRIDGETOWN
90 k

Saint Patricks

The Crane

Sargeant

Charnocks

Saint Lawrence

Providence

ATLANTIC
OCEAN

CARIBBEAN
SEA

0 _____ 6 miles

0 _____ 6 km

South America

Physically dominated by the Andes Mountains along its western edge and low-lying plains and river basins to the east, South America boasts a huge range of environments, from tropical rainforests, through arid desert plateaus and extensive prairies, to chilly temperate climes. Geography has gifted the continent with rich reserves of minerals and fossil fuels, but also huge biodiversity; the Amazon rainforest alone harbours some 10 per cent of the world's known species, and at least as many unknown.

Human settlement from the north seems to have begun around 9,000 years ago. From the second millennium BCE, settled agriculture gave rise to advanced but poorly understood civilizations that left behind enigmatic traces (such as the famous Nazca lines), and ultimately produced the Inca Empire, which dominated the west of the continent until the early 16th century. The European colonization that began in the 1530s came primarily from Spain and Portugal. Today, the majority of South Americans live in coastal areas and the Andean highlands.

Peru

Peru has three main geographical areas: a coastal desert belt in the west and the Andean highlands and rainforest in the east. The Inca Empire, which ruled much of Andean South America, arose from the area that is modern-day Peru in 1438, and lasted until the Spanish conquest of 1533. In the aftermath, European diseases devastated the indigenous population, and many were forced to work in mines. Peru remained part of the Spanish empire until the early 19th century, when it joined with other colonies to fight for independence. This was finally achieved in 1824, although independent Peru alternated between elected and military rulers, and fought territorial wars with its neighbours.

Since 1980, more than 70,000 people have died as a result of internal conflict between government forces and communist guerrillas which, although diminished in scale since 2000, is ongoing. Peru has also struggled with economic problems, corruption and authoritarian government, but has grown more stable and democratic since the early 21st century.

Republic of Peru
12.05° S, 77.04° W
UTC –5

- ☐ 1,285,220 km²
  (496,226 sq mi)
- ○ Dry desert in west,
  tropical in east
- ✝ 31,773,839
- ⊡ 24.8/km²
  (64/sq mi)
- ↟↟ 1.3%
- ⬭ 78.9 : 21.1%
- ⋔ 45% Amerindian
  37% mestizo
  15% white,
  3% others
- ♟ Spanish, Quechua,
  Aymara
- ▯ Christianity
- ♨ Unitary
  semi-presidential
  republic
- ⦿ CAN, G24, G77, IMF,
  NAM, OAS, PA, UN,
  WB, WTO
- ⬐ Peruvian Sol (PEN)
- ▦ $192.09 bn
- ▧ $6,046
- ⬒ Mining (particularly
  silver and copper),
  agriculture

Ecuador

Prior to Spanish conquest in 1533, most of modern-day Ecuador was part of the Inca Empire that dominated western South America. Ecuador, at the time named Quito after its capital, won independence from Spain in 1822. Along with other newly independent states – Colombia, Venezuela, Ecuador and Panama – it joined the federation of Gran Colombia (leaving in 1830, a year before its collapse). As a fully independent republic, Ecuador was politically unstable, with a rapid turnover of rulers. From 1904–42, it lost territory to several neighbours. In addition, a series of territorial wars with Peru, going back to 1821, were only settled by a peace accord in 1998.

Oil was discovered in Ecuador in 1967, and the country remains highly reliant on the petroleum industry. Following a short period of rule by a military junta, democratic elections restarted in 1979, and have continued since. The country's environmentally diverse territory also includes the Galápagos Islands, located 1,000 km (600 miles) from the mainland in the Pacific Ocean.

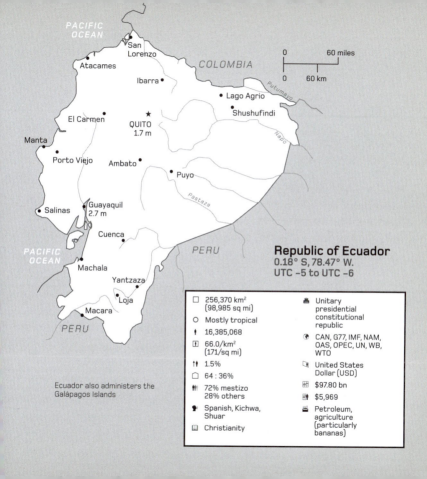

PACIFIC
OCEAN

San
Lorenzo

Atacames

COLOMBIA

Ibarra

Lago Agrio

Shushufindi

El Carmen

QUITO
1.7 m

Manta

Porto Viejo

Ambato

Puyo

Pastaza

Salinas

Guayaquil
2.7 m

Cuenca

PERU

Machala

Yantzaza

Loja

Macara

PERU

Putumayo

Napo

Ecuador also administers the
Galápagos Islands

Republic of Ecuador
0.18° S, 78.47° W.
UTC –5 to UTC –6

- ☐ 256,370 km²
  (98,985 sq mi)
- O Mostly tropical
- † 16,385,068
- ⊞ 66.0/km²
  (171/sq mi)
- †† 1.5%
- ☐ 64 : 36%
- ⋔ 72% mestizo
  28% others
- Spanish, Kichwa,
  Shuar
- ▢ Christianity

- ⚒ Unitary
  presidential
  constitutional
  republic
- ◉ CAN, G77, IMF, NAM,
  OAS, OPEC, UN, WB,
  WTO
- ▢ United States
  Dollar (USD)
- ▦ $97.80 bn
- ▤ $5,969
- ⚒ Petroleum,
  agriculture
  (particularly
  bananas)

Colombia

Most of Colombia's population lives in the Andean highlands, running roughly parallel to the Caribbean and Pacific coastal regions; lowland plains cover most of the eastern interior. Colombia features part of the Amazon rainforest. During the first half of the 16th century, the Spanish conquered Colombian territory, forcing the indigenous population to work for them; many died from violence or disease. Spain also imported slaves from Africa. In 1810, the people of Colombia rose up against colonial rule and, from 1819–30, the country joined Venezuela, Ecuador and Panama as part of Gran Colombia. Colombia then became a wholly independent republic, also incorporating Panama until 1903. Stability has often been elusive for Colombia, and violence virtually omnipresent. This has taken various forms: civil war, conflict related to organized crime and fighting in rural areas between left-wing guerrillas and right-wing paramilitaries. A major breakthrough was made in 2016, when the government signed a peace deal with the main guerrilla group, ending over five decades of armed insurgency.

Republic of
Colombia
4.71° N, 74.07° W.
UTC −5

☐	1,141,749 km² (440,832 sq mi)
○	Mostly tropical
✝	48,653,419
▦	43.9/km² (114/sq mi)
↑↑	0.9%
◠	76.7 : 23.3%
⋔	84% mestizo and white 10% Afro-Colombian 6% others
♟	Spanish
▢	Christianity
♠	Unitary presidential constitutional republic
◉	CAN, G24, G77, IMF, NAM, OAS, PA, UN, WB, WTO
⬚	Colombian Peso (COP)
▦	$282.46 bn
▦	$5,806
⛏	Mining (particularly gold, silver, emeralds and platinum), oil, agriculture (particularly coffee)

Chile

Running 4,200 kilometres (2,600 miles) down South America's western seaboard, Chile has a landscape of mountains, fjords and desert. The most numerous indigenous people are the Mapuche, who fought off the southward expansion of the Inca empire in the late 15th century. However, they could not prevent Spain conquering the country during the mid-16th century. Spanish rule ended in 1818; during the late 19th century, the independent republic subjugated the Mapuche in the south and expanded northwards following victory over Bolivia and Peru in the War of the Pacific (1879–83). Thanks to copper and nitrate mining, Chile grew more prosperous, and migration from Europe increased. Oligarchic rule slowly gave way to democracy, and in 1970 a left-wing president, Salvador Allende, was voted into office. Three years later a US-backed military coup overthrew him, bringing Augusto Pinochet to power with an authoritarian regime that killed and tortured thousands of opponents. Following the loss of a referendum on his position, Pinochet stepped down in 1990. Chile is now democratic, stable and increasingly prosperous.

Republic of Chile

33.44° S, 70.66° W.
UTC −3 and UTC −5

- ☐ 756,096 km²
  (291,930 sq mi)
- ○ Mainly temperate
- ✝ 17,909,754
- ▦ 24.1/km²
  (62/sq mi)
- ⇈ 0.8%
- ◻ 89.7 : 10.3%
- ⚥ 89% European
  11% others
- ♟ Spanish
- 📖 Christianity
- ⚒ Unitary
  presidential
  constitutional
  republic
- ◈ G-15, G77, IMF, NAM,
  OAS, OECD, PA, UN,
  WB, WTO
- ⌧ Chilean Peso (CLP)
- ▦ $247.03 bn
- ▦ $13,793
- ⛏ Mining (particularly
  copper), agriculture,
  forestry

'Insular Chile' is a collection of islands in the Pacific
Ocean made up of: the Juan Fernández Islands, the
Desventuradas Islands, Easter Island and Salas y
Gómez Island; the government also claims Chilean
Antarctic Territory

PERU

BOLIVIA

Arica
Pisagua
Iquique
Tocopilla   Chuquicamata

PACIFIC
OCEAN

Antofagasta
Taltal
Pueblo
Hundido
Caldera
Huasco

ISLA SAN
AMBROSIO

ISLA
SAN FELIX

La Serena

ANDES

SANTIAGO
6.5 m
San Antonio

Talca

ARGENTINA

ISLAS JUAN
FERNANDEZ

Cauquenes
Concepcion
Lebu

Temuco

Valdivia
Osorno
Puerto Montt

Castro

0   400 miles

0   400 km

Puerto Aisen

ARGENTINA

PACIFIC
OCEAN

SOUTH
ATLANTIC
OCEAN

Punta Arenas

ISLA DE LOS
ESTADOS

Brazil

Although Brazil is the largest nation in South America, its interior is dominated by the Amazon rainforest and so most of the population lives along its Atlantic coast. Indigenous Amerindians settled around 9000 BCE, but Portuguese rule, beginning in 1500, signalled a rapid decline of the local population and the introduction of African slaves to work on the colonial economic mainstays of sugar-cane plantations and gold mining. In 1822, the son of the Portuguese king declared himself ruler of the independent Empire of Brazil, but a federal republic replaced the monarchy in 1889. Today, there are 26 states and one federal district around Brasilia, the capital since 1960. Brazil's population grew rapidly owing to migration from Europe, and the country soon became the world's biggest coffee exporter. The military dominated Brazil until 1985, either ruling directly or through heavy influence on the country's leadership. Since then, Brazil has become a civilian-led democracy, and its economy has grown rapidly, thanks to its abundant natural resources and diverse industrial base.

Federative Republic of Brazil
15.79° S, 47.88° W
UTC −2 to UTC −5

VENEZUELA
COLUMBIA
GUYANA
SURINAME
FRENCH GUIANA FRANCE
Boa Vista
Macapa
Negro
Amazon
Manaus
Santarem
Belem
Sao Luis
NORTH ATLANTIC OCEAN
Altamira
Itaituba
Maraba
Fortaleza
Parnaiba
Benjamin Constant
Amazon
Purus
Madeira
Tapajos
Carajas
Teresina
Picos
Natal
Cruzeiro do Sul
Boca do Acre
Humaita
Porto Velho
Araguaina
Tocantins
Salgueiro
Recife
PERU
Guajara-Mirim
Juruena
Xingu
Palmas
Barreiras
Sao Francisco
Maceio
Aracaju
BOLIVIA
Caceres
Cuiaba
Araguaia
BRASILIA
4.2 m
Goiania
Ilheus
Rondonopolis
Uberlandia
Vitoria da Conquista
Corumba
Santa Fe do Sul
Grande
Belo Horizonte
BRAZILIAN HIGHLANDS
Campo Grande
Panorama
Sao Paulo
21.1 m
Vitoria
PARAGUAY
Parana
Curitiba
Rio de Janeiro
ARGENTINA
Sao Francisco do Sul
Florianopolis
Porto Alegre
Santa Maria
Rio Parana
SOUTH ATLANTIC OCEAN
URUGUAY
Rio Grande

0 — 400 miles
0 — 400 km

- ☐ 8,515,770 km² (3,287,957 sq mi)
- ○ Mostly tropical
- ♱ 207,652,865
- ⊞ 24.8/km² (64/sq mi)
- ⇧⇧ 0.8%
- ☐ 85.9 : 14.1%
- ♚ 47.7% white 43.1% mixed white and black 9.2% others
- ♟ Portuguese
- ☐ Christianity
- ♖ Federal presidential constitutional republic
- ☺ G-15, G24, G77, IMF, Mercosur, OAS, UN, WB, WTO
- ☐ Brazilian Real (BRL)
- ☐ $1.80 tn
- ☐ $8,650
- ☐ Transport equipment, mining, agriculture

Argentina

Although it includes mountains, jungles and arid steppe, the heart of the Argentinian landscape is the Pampas, a fertile expanse of grassy plains. Spanish rule over a region that was only sparsely populated by Amerindian peoples began in the early 16th century, and ended with independence in 1816 (following an 1810 revolution). Tensions between centrists in Buenos Aires and nationwide regionalists, however, led to civil wars that lasted until 1876. In the late 19th and early 20th centuries, mass migration from Europe brought rapid population growth. The election of president Juan Perón in 1946 was a watershed moment; he remained in office until 1955 (returning in 1973–74), launching a series of populist reforms. Perón's legacy remains a powerful force, but the armed forces have also been a major influence; military junta that took power in 1976, oversaw the deaths of thousands of opposition figures. An unsuccessful attempt to conquer the Falkland Islands resulted in the regime's overthrow in 1983; democratic civilian rule has been restored, but while Argentina is industrialized and prosperous, its economy has suffered numerous financial crises.

Argentine Republic
34.60° S, 58.38° W
UTC −3

☐	2,780,400 km² (1,073,518 sq mi)
○	Mostly temperate
⯅	43,847,430
⊞	16/km² (41/sq mi)
⭡⭡	1%
◠	91.9 : 8.1%
⭢⭠	97% European and mestizo 3% others
⚑	Spanish
▥	Christianity
⛭	Federal presidential constitutional republic
⊛	G-15, G24, G77, IMF, Mercosur, OAS, UN, WB, WTO
⬚	Argentine Peso (ARS)
▦	$545.87 bn
👤	$12,449
⛏	Agriculture, automotives, electronics

Venezuela

The Venezuelan landscape includes Andean mountains, Amazon jungle and tropical grassland, as well as Lake Maracaibo, the largest body of water in South America. Spanish colonization began in coastal areas during the early 16th century and gradually pushed inland, putting down resistance from the indigenous population. From 1810–23, Venezuelan nationalists fought to win freedom from Spain, joining Gran Colombia along with Ecuador, Colombia and Panama. The country became fully independent in 1830, and for over a century *caudillos* (local strongmen with private armies) and military dictators held sway. Democracy was only permanently established in 1959. The presidency of Hugo Chávez (1999–2013) saw a series of populist reforms, but central government became increasingly authoritarian. Venezuela had been largely agricultural until oil drilling began in 1914; petroleum exports have since been the backbone of the economy. Despite having the largest proven oil reserves in the world, corruption and mismanagement mean poverty and crime continue to be endemic. Since 2012, falling oil prices have led to economic crisis.

Bolivarian Republic of Venezuela
10.48° N, 66.90° W
UTC −4

☐ 912,050 km² (352,144 sq mi)	☐ 89 : 11%	⊕ G-15, G77, IMF, Mercosur (suspended since 2016), OAS (announced withdrawal 2017), NAM, OPEC, UN, WB, WTO
○ Mostly tropical; moderate in highlands	⚥ 66% Mestizo 20% European 10% African 4% others	
† 31,568,179	⚲ Spanish	🗎 Venezuelan Bolívar (VEF)
⊡ 35.8/km² (93/sq mi)	🕮 Christianity	🌐 $371.00 bn (2013)
†† 1.3%	⚒ Federal presidential constitutional republic	🖩 $12,237 (2013)
		🛢 Oil

Bolivia

In the 15th century, the Andean highlands that covered the west of Bolivia were part of the Inca Empire, while a mixture of tribes ruled the eastern lowlands. By 1538, however, Spain had established control over the entire region, which they named Charcas (after its capital city, now Sucre). Indigenous peoples like the Aymara and Quechua were conscripted to work in silver mines, under a system of forced resettlement and labour known as the *mit'a*. Independence followed a 16-year struggle against Spain, in 1825; the new republic was named Bolivia, after Simon Bolívar, the leading figure in the Spanish-American Wars of Independence. However, the country struggled; it was politically unstable and military losses to its neighbours reduced its territory by half, leaving Bolivia landlocked. Following a succession of coups, internal conflicts and military rulers, the country has been under democratic civilian control since 1982. Despite mineral wealth and reserves of natural gas, it remains the poorest country in South America, and many farmers rely on growing coca, which is used to make cocaine.

□ 1,088,580 km²
(424,164 sq mi)

○ Tropical to semi-arid

† 10,887,882

① 10.1/km²
(26/sq mi)

†† 1.5%

◠ 68.9 : 31.1%

♛ 68% mestizo (mixed white and Amerindian) 20% indigenous 12% others

♟ Spanish is most widely spoken although there are 36 other official languages

▢ Christianity

♜ Unitary presidential constitutional republic

◉ CAN, G77, IMF, NAM, OAS, UN, WB, WTO

⌨ Bolivian Boliviano (BOB)

▦ $33.81 bn

▨ $3,105

▱ Natural gas and mineral mining (particularly tin and zinc)

Plurinational State of Bolivia
16.50° S, 68.15° W (La Paz)
UTC −4

Bolivia has two capitals – Sucre (constitutional) and La Paz (legislative).

BRAZIL
Riberalta
Madre de Dios
PERU
Principe Da Beira
San Joaquin
Magdalena
BRAZIL
Itenes
Santa Ana
Rurrenabaque
San Ignacio
Trinidad
San Ignacio
Concepcion
Lake Titicaca
LA PAZ
1.8 m
Chulumani
San Ramon
San Miguel
Calacoto
Cochabamba
Okoruro
Totora
San Ignacio
Santa Cruz
de la Sierra
2.1 m
Oruro
Aiquile
Challapata
SUCRE
372 k
Potosi
PARAGUAY
Uyuni
Ticatica
A N D E S
Tupiza
Tarija
Villazon
CHILE
ARGENTINA

0   120 miles

0   120 km

Paraguay

The eponymous Paraguay River runs through the middle of this country, with arid plains to the west and grasslands, forests and mountains to the east. The main ethnic group is the Guaraní, whose language is still spoken by 90 per cent of Paraguayans. Spanish colonization began in 1537, but was sparse. Paraguay won independence in 1811, and José Gaspar Rodríguez de Francia, 'supreme dictator' from 1814–40, solidified the country's multicultural identity by forbidding people of European descent to marry each other. From 1864 to 1870, Paraguay was defeated in wars by Argentina, Brazil and Uruguay, losing two-thirds of its male population and half its territory. After decades of strife between liberals and conservatives, a military coup brought Alfredo Stroessner to power. His repressive dictatorship lasted until 1989, and the country's first free multi-party elections were not held until 1993. Paraguay has been generally stable since, but remains one of the poorest countries in South America, heavily reliant on agriculture and the export of hydroelectric power.

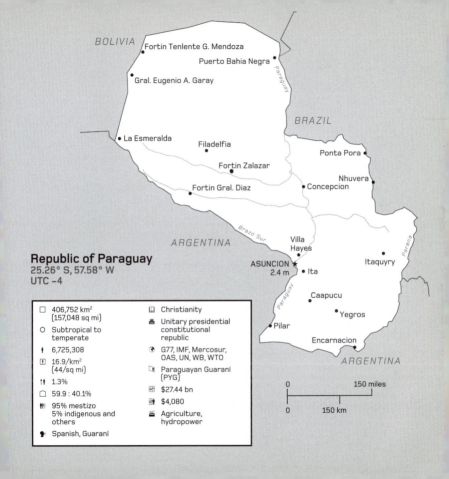

BOLIVIA

Fortin Tenlente G. Mendoza

Puerto Bahia Negra

Gral. Eugenio A. Garay

Paraguay

BRAZIL

La Esmeralda

Filadelfia

Fortin Zalazar

Ponta Pora

Nhuvera

Fortin Gral. Diaz

Concepcion

Brazo Sur

ARGENTINA

Villa Hayes

Parana

ASUNCION
2.4 m

Ita

Itaquyry

Caapucu

Yegros

Paraguay

Pilar

Encarnacion

ARGENTINA

Republic of Paraguay
25.26° S, 57.58° W
UTC −4

☐	406,752 km² (157,048 sq mi)	📖	Christianity
○	Subtropical to temperate	⚒	Unitary presidential constitutional republic
🕯	6,725,308	🌐	G77, IMF, Mercosur, OAS, UN, WB, WTO
⊡	16.9/km² (44/sq mi)	💱	Paraguayan Guaraní (PYG)
⚤	1.3%		$27.44 bn
☐	59.9 : 40.1%		$4,080
👥	95% mestizo 5% indigenous and others	⚒	Agriculture, hydropower
🗣	Spanish, Guaraní		

0 ——————— 150 miles

0 ——————— 150 km

Guyana

Geographically part of South America, but culturally Caribbean, Guyana is 75 per cent rainforest. Its indigenous peoples are Amerindian tribes who settled during the first millennium BCE. The Dutch set up trading posts from 1580 and moved inland, eventually establishing the colonies of Essequibo, Berbice and Demerara (named after local rivers). They imported African slaves to cultivate sugar cane, but faced attacks from the local population and other European powers before ceding control to Britain in 1814. The British merged the three existing colonies into one, British Guiana, in 1831. Following the abolition of slavery in 1834, indentured workers from the Indian subcontinent arrived in high numbers. From the mid-20th century, the colony moved towards self-rule, winning independence as Guyana in 1966. Long-term racial tension between the Afro- and Indo-Guyanese often leads to violence, particularly during elections, as support for the two main parties is principally organized along ethnic lines. The country also faces territorial disputes with Venezuela and Suriname, which have arisen from potential oil deposits.

VENEZUELA

ATLANTIC OCEAN

• Mabaruma

• Port Kaituma

Matthawa Ridge

• Charity

Cuyuni

Spring Garden

GEORGETOWN
124 k

• Arimu Mine

Peters Mine •

Bartica •

Rose Hall

New Amsterdam

• Rockstorie

Issano •

Linden

Corriverton

• Wismar
Kalkuni

Potaro Landing •

Takama

• Iluni

Mahdia •

• Orinduik

Kumpukari

SURINAME

0   60 miles

Surama •

0   60 km

Good Hope •

Essequibo

• Lethem

• Dadanawa

• Iserton

BRAZIL

**Co-operative
Republic
of Guyana**
6.80° N, 58.16° W
UTC –4

▢	214,970 km² (83,000 sq mi)
○	Tropical
♠	773,303
⊡	3.9/km² (10/sq mi)
⚢	0.6%
◠	28.7 : 71.3%
♛	40% East Indian 29% black 20% mixed 11% others
☙	English
▣	Christianity, Hinduism
♟	Unitary presidential republic
☯	CARICOM, CON, G77, IMF, NAM, OAS, UN, WB, WTO
▱	Guyanese Dollar (GYD)
▨	$3.45 bn
▨	$4,457
⛊	Agriculture, mining (gold and bauxite), forestry

Uruguay

A lowland landscape of rolling hills and plains hemmed by river deltas, Uruguay's small indigenous Charrúa population were overwhelmed in the 16th–18th centuries as the Spanish and Portuguese empires vied for control. Spain emerged victorious in 1776, but its rule came to an end in 1811 during the Spanish-American wars of independence. Initially, both Argentina and Brazil claimed Uruguay, and it did not win full independence until 1828.

Population grew rapidly in the 19th century, thanks to mass migration from Europe (mostly Spain and Italy); livestock farming contributed to growing prosperity. By the early 20th century, the government introduced social reforms that established South America's first welfare state. Uruguay has remained generally stable since that time, despite guerrilla warfare leading to violence and the armed forces seizing power in 1973. Civilian rule returned in 1985. With 80 per cent of its land used for agriculture, Uruguay remains a major exporter of beef and wool, with a robust economy that survived a major banking crisis in 2002.

ARGENTINA

Uruguay

Artigas •

Rivera •

• Salto

• Tacurembo

BRAZIL

Colon •
Paysandu •

Melo •

Rio Branco •

Negro

Mercedes •

Treinta Y Tres •

Parana

SOUTH
ATLANTIC
OCEAN

Castillos •

San Carlos •

ARGENTINA

★ MONTEVIDEO
1.7 m

0   100 miles
0   100 km

**Oriental
Republic of
Uruguay**
34.90° S, 56.16° W
UTC −3

☐ 176,220 km² (68,039 sq mi)	⚥ 88% white 8% mestizo 4% others	⊕ G77, IMF, Mercosur, OAS, UN, WB, WTO
○ Warm temperate	⚲ Spanish	⌨ Uruguayan Peso (UYU)
✦ 3,444,006	⌂ Christianity	▦ $52.42 bn
⊞ 19.7/km² (51/sq mi)	⚒ Unitary presidential constitutional republic	▨ $15,221
↟↟ 0.4%		⛏ Agriculture
◠ 95.5 : 4.5%		

Suriname

With four-fifths of its southern territory covered by tropical rainforest, Suriname has the lowest population density in South America. Its indigenous population resisted European settlers during the first half of the 17th century, and it was not until 1651 that a successful colony was founded by the English. In 1667, it was captured by the Netherlands, who established sugar-cane and coffee plantations; they also transported African slaves there to work for them. When slavery was abolished in 1863, indentured workers from India, Madeira, Java and China arrived. Independence came in 1975; prior to this, one-third of the population immigrated to the Netherlands, fearing instability when Dutch rule ended. Their concerns were well founded, as Surinamese politics have been riven by corruption and ethnic-based rivalry. Military rule dominated the 1980s and, from 1986–92, a civil war raged between the government and rebel guerrillas in the interior. Democracy has largely been restored, although the economy's long-term reliance on mineral mining (particularly of bauxite) makes it vulnerable to global price changes.

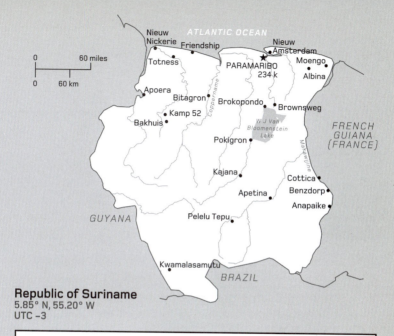

ATLANTIC OCEAN

Nieuw
Nickerie • Friendship
Nieuw
Amsterdam
Moengo
Totness
★ PARAMARIBO
234 k
Albina
Apoera
Bitagron
Brokopondo • Brownsweg
Kamp 52
W J Van
Bloomenstein
Lake
Bakhuis
FRENCH
GUIANA
(FRANCE)
Pokigron
Kajana
Cottica
Apetina
Benzdorp
Anapaike
Pelelu Tepu
GUYANA
Kwamalasamutu
BRAZIL

0 — 60 miles
0 — 60 km

Coppername
Marowijne

Republic of Suriname
5.85° N, 55.20° W
UTC –3

☐ 163,820 km² (63,251 sq mi)	♔ 27% Indian 22% Maroon 16% Creole 14% Javanese 21% others	⊕ CARICOM, G77, IMF, NAM, OAS, UN, WB, WTO
○ Tropical		
♀ 558,368	♟ Dutch	▣ Surinamese Dollar (SRD)
▣ 3.6/km² (9/sq mi)	▥ Christianity	▦ $3.62 bn
♂ 0.9%	♨ Unitary parliamentary republic	▧ $6,484
◻ 66 : 34%		▩ Oil, mining (particularly gold and bauxite)

Index of Countries

Afghanistan 256
Albania 74
Algeria 126
Andorra 24
Angola 156
Antigua and Barbuda 376
Argentina 400
Armenia 234
Australia 310
Austria 46
Azerbaijan 238
Bahamas, The 364
Bahrain 244
Bangladesh 276
Barbados 386
Belarus 94
Belgium 26
Belize 352
Benin 138
Bhutan 278
Bolivia 404
Bosnia and Herzegovina 64
Botswana 168
Brazil 398
Brunei 296
Bulgaria 92
Burkina Faso 130
Burundi 184
Cambodia 292
Cameroon 150
Canada 344
Cape Verde 106

Central African Republic 164
Chad 162
Chile 396
China 268
Colombia 394
Comoros 206
Costa Rica 358
Côte d'Ivoire 128
Croatia 58
Cuba 360
Cyprus 102
Czech Republic 50
Democratic People's
    Republic of Korea 302
Democratic Republic of the
    Congo 160
Denmark 42
Djibouti 204
Dominica 380
Dominican Republic 370
Ecuador 392
Egypt 176
El Salvador 350
Equatorial Guinea 142
Eritrea 200
Estonia 96
Ethiopia 196
Federated States of
    Micronesia 318
Fiji 334
Finland 84
France 22

Gabon 148
Gambia, The 112
Georgia 230
Germany 34
Ghana 132
Greece 78
Grenada 378
Guatemala 348
Guinea 116
Guinea-Bissau 114
Guyana 408
Haiti 368
Honduras 354
Hungary 68
Iceland 12
India 262
Indonesia 282
Iran 236
Iraq 228
Ireland 14
Israel 218
Italy 38
Jamaica 366
Japan 306
Jordan 222
Kazakhstan 242
Kenya 198
Kiribati 330
Kuwait 240
Kyrgyzstan 264
Laos 288
Latvia 88

Lebanon 224
Lesotho 180
Liberia 124
Libya 152
Liechtenstein 44
Lithuania 86
Luxembourg 32
Macedonia 82
Madagascar 208
Malawi 194
Malaysia 286
Maldives 266
Mali 122
Malta 62
Marshall Islands 324
Mauritania 110
Mauritius 212
Mexico 346
Moldova 100
Monaco 40
Mongolia 274
Montenegro 70
Morocco 120
Mozambique 190
Myanmar 280
Namibia 158
Nauru 328
Nepal 272
Netherlands 28
New Zealand 312
Nicaragua 356
Niger 136
Nigeria 140
Norway 30
Oman 250
Pakistan 258
Palau 316

Palestine 216
Panama 362
Papua New Guinea 320
Paraguay 406
Peru 390
Philippines 298
Poland 60
Portugal 16
Qatar 246
Republic of Korea 304
Republic of the Congo 154
Romania 80
Russia 76
Rwanda 182
Samoa 336
San Marino 52
São Tomé and Príncipe 144
Saudi Arabia 220
Senegal 108
Serbia 72
Seychelles 210
Sierra Leone 118
Singapore 294
Slovakia 66
Slovenia 56
Solomon Islands 322
Somalia 202
South Africa 166
South Sudan 174
Spain 18
Sri Lanka 270
St Kitts and Nevis 372
St Lucia 384
St Vincent and the
   Grenadines 382
Sudan 170
Suriname 412

Swaziland 192
Sweden 48
Switzerland 36
Syria 226
Tajikistan 260
Tanzania 186
Thailand 284
Timor-Leste 300
Togo 134
Tonga 338
Trinidad and Tobago 374
Tunisia 146
Turkey 98
Turkmenistan 252
Tuvalu 332
Uganda 188
Ukraine 90
United Arab Emirates 248
United Kingdom 20
United States of America
   342
Uruguay 410
Uzbekistan 254
Vanuatu 326
Vatican City 54
Venezuela 402
Vietnam 290
Yemen 232
Zambia 172
Zimbabwe 178

First published in the United States by Quercus in 2018

Hachette Book Group
1290 Avenue of the Americas
New York, NY 10104

A Hachette UK Company

Copyright © Quercus 2018
Text by Jacob F. Field

Design and editorial by Pikaia Imaging
Edited by Anna Southgate
Editorial assistance: Suzanne Elliot, Kaleesha Williams

Library of Congress Control Number: 2018949430

PB ISBN 9781635061154
EBOOK ISBN 9781635061161

10 9 8 7 6 5 4 3 2 1

Printed and bound in China

Acknowledgements

Maps in this book were compiled using resources from:
www.mapresources.com and Mountain High Maps (www.digiwis.com)

Data boxes compiled using information from the CIA World Fact Book:
https://www.cia.gov/library/publications/the-world-factbook/